Claes Johnson

Numerical solution of partial differential equations by the finite element method

Library of Congress cataloging in publication data available

British Library cataloging in publication data available

4. printing
Printed in Sweden
Studentlitteratur, Lund, 1992

Published in Sweden, Finland, Norway, Denmark & Iceland by Studentlitteratur, Lund.
Published elsewhere by the Press Syndicate of the University of Cambridge
The Pitt Building, Trumpington Street, Cambridge CB2 1RP
32 East 57th Street, New York, NY 10022, USA
10 Stamford Road, Oakleigh, Melbourne 3166, Australia
ISBN 0 521 345 146 (hardcovers)
0 521 347 580 (paperback)

Contents

Preface

The purpose of this book is to give an easily accessible introduction to the finite element method as a general method for the numerical solution of partial differential equations in mechanics and physics covering all the three main types of equations, namely elliptic, parabolic and hyperbolic equations. The main part of the text is concerned with linear problems, but a chapter indicating extensions to some nonlinear problems is also included. There is also a chapter on finite element methods for integral equations connected with elliptic problems. The book is based on material that I have used in undergraduate courses at Chalmers University of Technology, Göteborg. The first half of the book (Chapters 1–7), which treat elliptic problems in a rather standard way, is a translation of a textbook in Swedish that appeared 1981 [J1]. Two chapters on parabolic and hyperbolic problems present recent developments based on my work on discontinuous Galerkin and streamline diffusion type finite element methods using, in particular, finite elements for the time discretization as well. For first order hyperbolic problems these are the first finite element methods with satisfactory properties and thus show promise of extensive applications. For parabolic problems, time-discretization by the discontinuous Galerkin method gives new efficient methods and makes a precise error analysis with associated automatic time step control possible for the first time.

The emphasis of the text is on mathematical and numerical aspects of the finite element method but many applications to important problems in mechanics and physics are also given. I have tried to keep the mathematics as simple as possible while still presenting significant results and maintaining a natural mathematical framework. Lately I have used the text of the book as part of the material in a series of undergraduate courses on partial differential equations leading up to graduate level treating in integrated form both mathematical questions on existence and regularity together with numerical methods. I have found this to be a fruitful approach where, on one hand the numerical methods can be given the necessary mathematical background, and on the other hand, the fascinating and important techniques of solving differential equations numerically using computers can give crucial motivation for the theoretical mathematical studies. In fact, the numerical and mathematical aspects are intimately connected and the new possibilities of computer simulation makes proper understanding of the mathematical structure and properties of the mathematical models very important also in applications. In the present book only a bare minimum of mathematical background is included and the reader is referred to the literature for a more complete account.

The list of references is limited and many contributions important for the development of the various subjects have been omitted. I have given just a few references leading into the very rich literature on finite element methods.

I want to thank Prof Raymond Chandler for revising the English, Tekn Lic Peter Hansbo for supplying most of the numerical results and Dr Kenneth Eriksson for reading parts of the material. Special thanks also to Mrs Yumi Karlsson who swiftly typed a first version of the text and with great patience helped me with seemingly endless alterations and corrections.

Göteborg in July 1987

Claes Johnson

0. Introduction

0.1 Background

The mathematical models of science and engineering mainly take the form of differential or integral equations. With the rapid development of high speed computers over the last decades the possibilities of efficiently utilizing these models have dramatically increased. Using computer-implemented mathematical models, one can simulate and analyze complicated systems in engineering and science. This reduces the need for expensive and time-consuming experimental testing and makes it possible to compare many different alternatives for optimization, etc. In fact, with the new possibilities an intense activity has started in Computer Aided Design, Engineering and Manufacturing (CAD, CAE and CAM) which is bringing revolutionary changes to engineering science and practice, and a new scientific field "scientific computing" is emerging as a complement to theoretical and experimental science.

To use mathematical models on a computer one needs numerical methods. Only in the very simplest cases is it possible to find exact analytical solutions of the equations in the model, and in general one has to rely on numerical techniques for finding approximate solutions. The *finite element method* (FEM) is a general technique for numerical solution of differential and integral equations in science and engineering. The method was introduced by engineers in the late 50's and early 60's for the numerical solution of partial differential equations in structural engineering (elasticity equations, plate equations, etc). At this point the method was thought of as a generalization of earlier methods in structural engineering for beams, frames and plates, where the structure was subdivided into small parts, so-called finite elements, with known simple behaviour. When the mathematical study of the finite element method started in the mid 60's it soon became clear that in fact the method is a general technique for numerical solution of partial differential equations with roots in the variational methods in mathematics introduced in the beginning of the century. During the 60's and 70's the method was developed, by engineers, mathematicians and numerical analysts, into a general method for numerical solution of partial differential equations and integral equations with applications in many areas of science and engineering. Today, finite element methods are used extensively (often integrated in CAD or CAE-systems) for problems in structural engineering, strength of materials, fluid mechanics, nuclear engineering, electro-magnetism, wave-propagation, scattering, heat conduction, convection-diffusion processes, integrated circuits, petroleum engineering, reaction-diffusion processes and many other areas.

0.2 Difference methods – Finite element methods

The basic idea in any numerical method for a differential equation is to *discretize* the given continuous problem with infinitely many degrees of freedom to obtain a *discrete problem* or system of equations with only finitely many unknowns that may be solved using a computer. The classical numerical method for partial differential equations is the *difference method* where the discrete problem is obtained by replacing derivatives with difference quotients involving the values of the unknown at certain (finitely many) points.

The discretization process using a finite element method is different. In this case we start from a reformulation of the given differential equation as an equivalent *variational problem*. In the case of elliptic equations this variational problem in basic cases is a minimization problem of the form

(*M*) Find $u \in V$ such that $F(u) \leq F(v)$ for all $v \in V$,

where V is a given set of admissible functions and $F: V \rightarrow R$ is a *functional* (i e $F(v) \in R$ for all $v \in V$ with R denoting the set of real numbers). The functions v in V often represent a continuously varying quantity such as a displacement in an elastic body, a temperature, etc, F(v) is the total energy associated with v and (*M*) corresponds to an equivalent characterization of the solution of the differential equation as the function in V that minimizes the total energy of the considered system. In general the dimension of V is infinite (i e the functions in V cannot be described by a finite number of parameters) and thus in general the problem (*M*) cannot be solved exactly. To obtain a problem that can be solved on a computer the idea in the finite element method is to replace V by a set V_h consisting of simple functions only depending on finitely many parameters. This leads to a finite-dimensional minimization problem of the form:

(M_h) Find $u_h \in V_h$ such that $F(u_h) \leq F(v)$ for all $v \in V_h$.

This problem is equivalent to a (large) linear or nonlinear system of equations. The hope is now that the solution u_h of this problem is a sufficiently good approximation of the solution u of the original minimization problem (M), i e, the original partial differential equation. Usually one chooses V_h to be a subset of V (in other words $V_h \subset V$, i e, if $v \in V_h$ then $v \in V$) and in this case (M_h) corresponds to the classical *Ritz-Galerkin method* that goes back to the beginning of the century. The special feature of a finite element method as a particular Ritz-Galerkin method is the fact that the functions in V_h are chosen to be *piecewise polynomial.* As will be seen below, one may also start from more general variational formulations than the minimization problem (M); this corresponds e g to so-called *Galerkin methods.*

To solve a given differential or integral equation approximately using the finite element method, one has to go through basically the following steps:

(i) variational formulation of the given problem,
(ii) discretization using FEM: construction of the finite dimensional space V_h,
(iii) solution of the discrete problem,
(iv) implementation of the method on a computer: programming.

Often there are several different variational formulations that may be used depending e g on the choice of dependent variables. The choice of finite dimensional subspace V_h, i e, essentially the choice of the finite element discretization or the finite elements, is influenced by the variational formulation, accuracy requirements, regularity properties of the exact solution etc. To solve the discrete problem one needs optimization algorithms and/or methods for the numerical solution of large linear or nonlinear systems of equations. In this book we shall consider all the steps (i)−(iv) with (iv) kept at an introductory level.

The advantage of finite element methods as compared with finite difference methods is that complicated geometry, general boundary conditions and variable or non-linear material properties can be handled relatively easily. In all these cases one meets unneccessary artificial complications with finite difference methodology. Further, the clear structure and versatility of the finite element method makes it possible to construct general purpose software for applications and there is also a large number of more or less general finite element codes available. Also, the finite element method has a solid theoretical foundation which gives added reliability and in many cases makes it possible to mathematically analyze and estimate the error in the approximate finite element solution.

0.3 Scope of the book

The purpose of this book is to give an introduction to the finite element method as a general technique for the numerical solution of partial differential and integral equations in science and engineering. The focus is on mathematical and numerical properties of the method, but we also consider many important applications to problems from various areas. An effort has been made to keep the mathematics simple while still presenting significant results and considering non-trivial problems of practical interest.

We will consider the three main types of partical differential equations, i e, elliptic, parabolic and hyperbolic equations. To connect these types of equations with problems in mechanics and physics, we recall that elliptic equations model for example static problems in elasticity, parabolic equations model time-dependent diffusion dominated processes, and hyperbolic equations are used to describe convection or wave-propagation processes. We also give an introduction to finite element methods for boundary integral equations associated with elliptic boundary value problems in mechanics and physics. These methods are referred to as *boundary element methods* (BEM). We will mainly consider linear problems and only briefly comment on some non-linear ones.

The material presented concerning elliptic problems is by now standard, but for parabolic and hyperbolic problems we present recent developments that have not earlier appeared in text books. With these later contributions it is possible to give a unified treatment of the three main types of partial differential equations as well as boundary integral equations. In all cases we emphasize the basic role played by the stability properties of the finite element method and the relation to the corresponding properties of the partial differential or integral equation.

The book is an extended version of an earlier book in Swedish by the author that has been used for several years in undergraduate courses for engineering students at Chalmers University of Technology, Göteborg, Sweden, and also at other Scandinavian universities.

The necessary prerequisites are relatively moderate: Basic courses in advanced calculus and linear algebra and preferably some acquaintance with the most well-known linear partial differential equations in mechanics and physics, such as the Poisson equation, the heat equation and the wave equation. With some oversimplification we may say that the mathematical tools used in the book reduce to the following: Green's formula, Cauchy's inequality and elementary calculus and linear algebra.

The problem sections play an important role in the presentation and the reader is urged to spend time to solve the problems.

As a general reference giving a more detailed presentation of the material in Chapters 1 to 5, we refer to [Ci] (see also [SF]). For variational methods for partial differential equations in mechanics and physics, see e g [DL], [ET] and [Ne].

1. Introduction to FEM for elliptic problems

In this chapter we introduce FEM for some elliptic model problems and study the basic properties of the method. We first consider a simple one-dimensional problem and then some two-dimensional generalizations.

1.1 Variational formulation of a one-dimensional model problem

Let us consider the two-point boundary value problem

$$(D) \qquad \begin{aligned} -u''(x) &= f(x) \quad \text{for } 0<x<1, \\ u(0) &= u(1)=0, \end{aligned}$$

where $v' = \frac{dv}{dx}$ and f is a given continuous function. By integrating the equation $-u''=f$ twice, it is easy to see that this problem has a unique solution u. We recall that the boundary value problem (D) can be viewed as modelling, in particular, the following situations in continuum mechanics:

A An elastic bar

Consider an elastic bar fixed at both ends subject to a tangential load of intensity f(x) (see Fig 1.1). Let $\sigma(x)$ and u(x) be the traction and tangential displacement at x, respectively, under the load f. Under the assumption of small displacements and a linearly elastic material, we have in the interval (0,1)

$$\begin{aligned} \sigma &= Eu' && \text{(Hooke's law)}, \\ -\sigma' &= f && \text{(equilibrium equation)}, \\ u(0) &= u(1)=0 && \text{(boundary conditions)}, \end{aligned}$$

where E is the modulus of elasticity. If we take here E=1 and eliminate σ, we obtain (*D*).

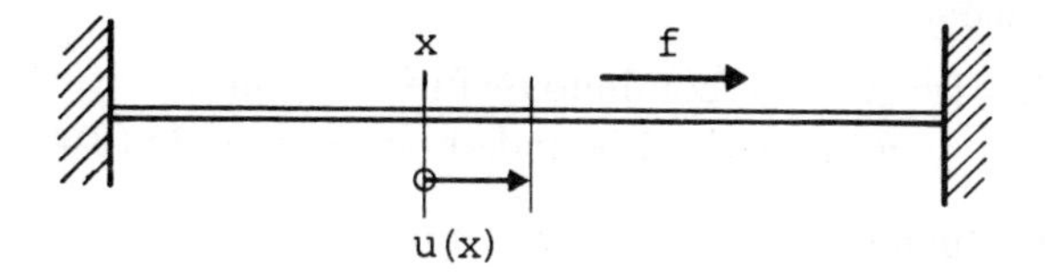

Fig 1.1

B An elastic cord

Consider an elastic cord with tension 1, fixed at both ends and subject to transversal load of intensity f (see Fig 1.2). Assuming again small displacements, we have that the transversal displacement u satisfies (*D*) (cf Problem 1.2).

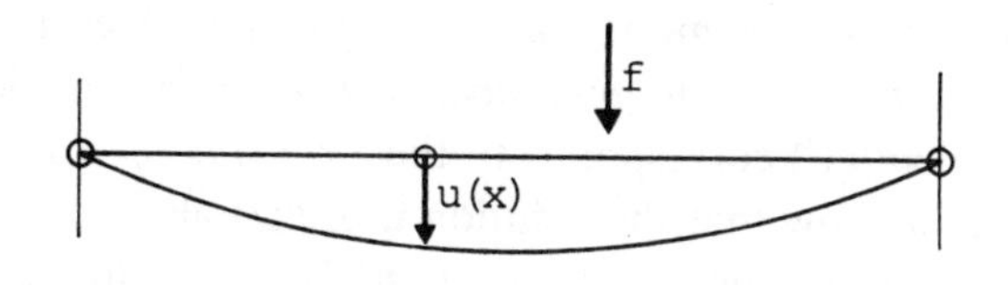

Fig 1.2

C Heat conduction

Let u be the temperature and q the heat flow in a heat conducting bar, subject to a distributed heat source of intensity f. Assuming the temperature to be zero at the end points, we have in the stationary case

$$\begin{aligned} -q &= ku' && \text{(Fourier's law)},\\ q' &= f && \text{(conservation of energy)},\\ u(0) &= u(1) = 0, \end{aligned}$$

where k is the heat conductivity, which again gives (*D*) if k=1.

We shall now show that the solution u of the boundary value problem or differential equation (*D*) also is the solution of a minimization problem (*M*) and a variational problem (*V*). To formulate the problems (*M*) and (*V*) we introduce the notation

$$(v,w)=\int_0^1 v(x)w(x)dx,$$

for real-valued piecewise continuous bounded functions. We also introduce the linear space

$$V=\{v: v \text{ is a continuous function on } [0,1],\ v' \text{ is piecewise continuous and bounded on } [0,1], \text{ and } v(0)=v(1)=0\},$$

and the linear functional $F: V \rightarrow R$ given by

$$F(v)=\frac{1}{2}(v', v')-(f, v).$$

The problems (*M*) and (*V*) are the following:

(*M*) Find $u \in V$ such that $F(u) \leqslant F(v) \quad \forall v \in V$,

(*V*) Find $u \in V$ such that $(u', v')=(f, v) \quad \forall v \in V$.

Let us notice that in the context of the problems A and B above, the quantity $F(v)$ represents the *total potential energy* associated with the displacement $v \in V$. The term $\frac{1}{2}(v', v')$ represents the internal elastic energy and (f,v) the load potential. Thus, the minimization problem (*M*) corresponds to the fundamental *Principle of minimum potential energy* in mechanics. Further the variational problem (*V*) corresponds to the *Principle of virtual work.*

Let us now first show that the solution u of (*D*) also is a solution of (*V*). To see this we multiply the equation $-u''=f$ by an arbitrary function $v \in V$, a so-called *test funtion* v, and integrate over the interval (0, 1) which gives

$$-(u'', v)=(f, v).$$

We now integrate the left-hand side by parts using the fact that $v(0)=v(1)=0$ to get

$$-(u'', v)=-u'(1)v(1)+u'(0)v(0)+(u', v')=(u', v'),$$

and we conclude that

$$(u', v')=(f, v) \qquad \forall v \in V, \tag{1.1}$$

which shows that u is a solution of (*V*).

Next, we show that the problems (*V*) and (*M*) have the same solutions. Suppose then first that u is a solution to (*V*), let $v \in V$ and set $w=v-u$ so that $v=u+w$ and $w \in V$. We have

$$F(v)=F(u+w)=\frac{1}{2}(u'+w', u'+w')-(f, u+w)$$

$$=\frac{1}{2}(u', u')-(f, u)+(u', w')-(f, w)+\frac{1}{2}(w', w') \geqslant F(u),$$

since by (1.1), $(u', w')-(f, w)=0$ and $(w', w')\geq 0$, which shows that u is a solution of (*M*). On the other hand, if u is a solution of (*M*) then we have for any $v\in V$ and real number ε

$$F(u)\leq F(u+\varepsilon v),$$

since $u+\varepsilon v\in V$. Thus, the differentiable function

$$g(\varepsilon)\equiv F(u+\varepsilon v)=\frac{1}{2}(u', u')+\varepsilon(u', v')+\frac{\varepsilon^2}{2}(v', v')-(f, u)-\varepsilon(f, v),$$

has a minimum at $\varepsilon=0$ and hence $g'(0)=0$. But

$$g'(0)=(u', v')-(f, v),$$

and we see that u is a solution of (*V*).

Let us also show that a solution to (*V*) is uniquely determined. Suppose then that u_1 and u_2 are solutions of (*V*), i e, $u_1, u_2\in V$ and

$$(u_1', v')=(f, v) \quad \forall v\in V,$$
$$(u_2', v')=(f, v) \quad \forall v\in V.$$

Subtracting these equations and choosing $v=u_1-u_2\in V$, we get

$$\int_0^1 (u_1'-u_2')^2\, dx=0,$$

which shows that

$$u_1'(x)-u_2'(x)=(u_1-u_2)'(x)=0 \qquad \forall x\in[0, 1].$$

It follows that $(u_1-u_2)(x)$ is constant on [0,1] which together with the boundary condition $u_1(0)=u_2(0)=0$ gives $u_1(x)=u_2(x)$, $\forall x\in[0,1]$, and the uniqueness follows.

To sum up, we have shown that if u is the solution to (*D*), then u is the solution to the equivalent problems (*M*) and (*V*) which we write symbolically as

$$(D)\Rightarrow(V)\Leftrightarrow(M).$$

Let us finally also indicate how to see that if u is the solution of (*V*) then u also satisfies (*D*). Thus, we assume that $u\in V$ satisfies

$$\int_0^1 u'v'dx-\int_0^1 fvdx=0 \qquad \forall v\in V.$$

If we now assume in addition that u'' exists and is continuous, then we can integrate the first term by parts to get, using the fact that $v(0)=v(1)=0$,

$$-\int_0^1 (u''+f)v\,dx=0 \qquad \forall v\in V.$$

But with the assumption that $(u''+f)$ is continuous this relation can only hold if (cf Problem 1.1)

$$(u''+f)(x)=0 \qquad 0<x<1,$$

and it follows that u is the solution of (D).

Thus we have seen that if u is the solution of (V) and in addition satisfies a regularity assumpton (u'' is continuous), then u is the solution of (D). It is now possible to show that if u is the solution of (V), then u in fact satisfies the desired regularity assumption and thus we have $(V)\Rightarrow(D)$ which shows that the three problems (D), (V) and (M) are equivalent (cf Section 1.5 below).

Problems

1.1 Show that if w is continuous on [0, 1] and

$$\int_0^1 wv\,dx=0 \qquad \forall v\in V,$$

then $w(x)=0$ for $x\in[0, 1]$.

1.2 Show that under suitable assumptions the problem B above can be given the formulation (1.1).

1.2 FEM for the model problem with piecewise linear functions

We shall now construct a finite-dimensional subspace V_h of the space V defined above consisting of piecewise linear functions. To this end let $0=x_0<x_1 \ldots <x_M<x_{M+1}=1$, be a partition of the interval (0,1) into subintervals $I_j=(x_{j-1}, x_j)$ of length $h_j=x_j-x_{j-1}$, $j=1, \ldots, M+1$ and set $h=\max h_j$. The quantity h is then a measure of how fine the partition is. We now let V_h be the set of functions v such that v is linear on each subinterval I_j, v is continuous on [0,1] and $v(0)=v(1)=0$ (cf Fig 1.3).

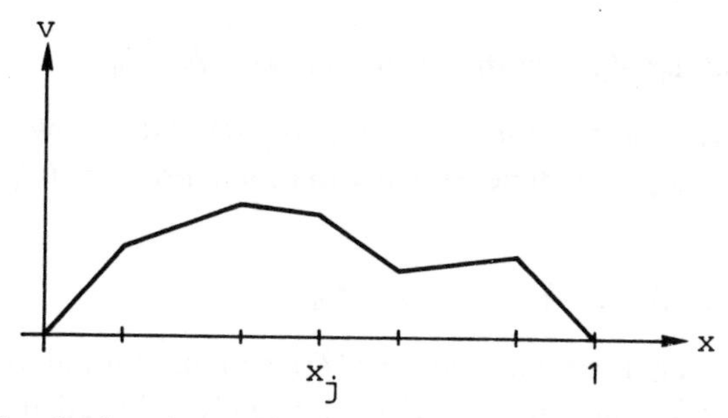

Fig 1.3 Example of a function $v \in V_h$

We observe that $V_h \subset V$. As parameters to describe a function $v \in V_h$ we may choose the values $\eta_j = v(x_j)$ at the node points x_j, $j=0,\ldots, M+1$. Let us introduce the *basis functions* $\varphi_j \in V_h$, $j=1,\ldots, M$, defined by

$$\varphi_j(x_i) = \begin{cases} 1 & \text{if } i=j \\ 0 & \text{if } i \neq j, \; i, j=1,\ldots, M, \end{cases}$$

i e, φ_j is the continuous piecewise linear function that takes the value 1 at node point x_j and the value 0 at other node points (see Fig 1.4).

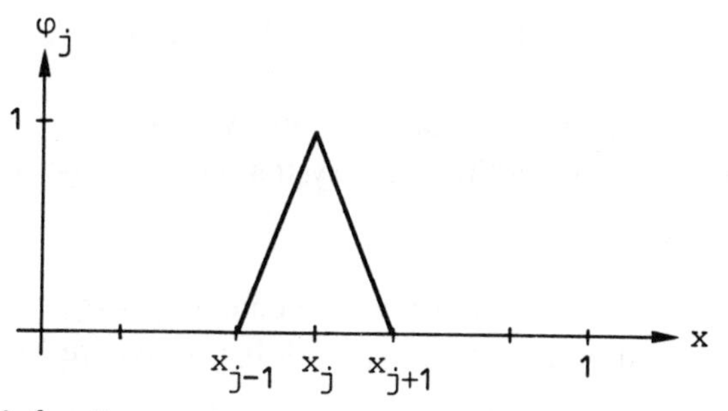

Fig 1.4 The basis function φ_j

A function $v \in V_h$ then has the representation

$$v(x) = \sum_{i=1}^{M} \eta_i \varphi_i(x), \qquad x \in [0,1],$$

where $\eta_i = v(x_i)$, i e, each $v \in V_h$ can be written in a unique way as a *linear combination* of the basis functions φ_i. In particular it follows that V_h is a *linear space of dimension* M with *basis* $\{\varphi_i\}_{i=1}^{M}$.

The finite element method for the boundary value problem (*D*) can now be formulated as follows:

(M_h) Find $u_h \in V_h$ such that $F(u_h) \leq F(v) \quad \forall v \in V_h$.

In the same way as above for the problems (M) and (V), we see that (M_h) is equivalent to the finite-dimensional variational problem (V_h): Find $u_h \in V_h$ such that

$$(1.2) \qquad (u_h', v') = (f, v) \qquad \forall v \in V_h.$$

Thus the finite element method for (D) can be formulated as (V_h) or equivalently (M_h). The problem (V_h) is usually referred to as *Galerkin's method* and (M_h) as *Ritz' method*. We observe that if $u_h \in V_h$ satisfies (1.2), then in particular

$$(1.3) \qquad (u_h', \varphi_j') = (f, \varphi_j) \qquad j = 1, \ldots, M,$$

and if these equations hold, then by taking linear combinations, we see that u_h satisfies (1.2). Since

$$u_h(x) = \sum_{i=1}^{M} \xi_i \varphi_i(x), \qquad \xi_i = u_h(x_i),$$

we can write (1.3)

$$(1.4) \qquad \sum_{i=1}^{M} \xi_i (\varphi_i', \varphi_j') = (f, \varphi_j) \qquad j = 1, \ldots, M,$$

which is a linear system of equations with M equations in M unknowns $\xi_1, \ldots, \xi_M$. In matrix form the linear system (1.4) can be written as

$$(1.5) \qquad A\xi = b,$$

where $A = (a_{ij})$ is the $M \times M$ matrix with elements $a_{ij} = (\varphi_i', \varphi_j')$, and where $\xi = (\xi_1, \ldots, \xi_M)$ and $b = (b_1, \ldots, b_M)$ with $b_i = (f, \varphi_i)$ are M-vectors:

$$A = \begin{bmatrix} a_{11} & \cdots & a_{1M} \\ \vdots & & \vdots \\ a_{M1} & \cdots & a_{MM} \end{bmatrix}, \quad \xi = \begin{bmatrix} \xi_1 \\ \cdot \\ \xi_M \end{bmatrix}, \quad b = \begin{bmatrix} b_1 \\ \vdots \\ b_M \end{bmatrix}.$$

The matrix A is called the *stiffness matrix* and b the *load vector*, with terminology from early applications of FEM in structural mechanics.

The elements $a_{ij} = (\varphi_i', \varphi_j')$ in the stiffness matrix A can easily be computed: We first observe that $(\varphi_i', \varphi_j') = 0$ if $|i-j| > 1$ since in this case for all $x \in [0,1]$ either $\varphi_i(x)$ or $\varphi_j(x)$ is equal to zero. Thus, the matrix A is tri-diagonal, i e, only the elements in the main diagonal and the two adjoining diagonals may be different from zero. We have for $j = 1, \ldots, M$,

$$(\varphi_j', \varphi_j') = \int_{x_{j-1}}^{x_j} \frac{1}{h_j^2}\,dx + \int_{x_j}^{x_{j+1}} \frac{1}{h_{j+1}^2}\,dx = \frac{1}{h_j} + \frac{1}{h_{j+1}},$$

and for $j=2, \ldots, M$,

$$(\varphi_j', \varphi_{j-1}') = (\varphi_{j-1}', \varphi_j') = -\int_{x_{j-1}}^{x_j} \frac{1}{h_j^2}\,dx = -\frac{1}{h_j}.$$

Note also that the matrix A is *symmetric* and *positive definite* since $(\varphi_i', \varphi_j') = (\varphi_j', \varphi_i')$ and with $v(x) = \sum_{j=1}^{M} \eta_j \varphi_j(x)$, we have

$$\sum_{i,j=1}^{M} \eta_i (\varphi_i', \varphi_j') \eta_j = \Big(\sum_{i=1}^{M} \eta_i \varphi_i', \sum_{j=1}^{M} \eta_j \varphi_j'\Big) = (v', v') \geq 0,$$

with equality only if $v' \equiv 0$, that is since $v(0)=0$ only if $v \equiv 0$, or $\eta_j = 0$ for $j=1, \ldots, M$. We recall that a symmetric $M \times M$ matrix $S = (s_{ij})$ is said to be positive definite if

$$\eta \cdot S\eta = \sum_{i,j=1}^{M} \eta_i s_{ij} \eta_j > 0 \qquad \forall \eta \in R^M,\ \eta \neq 0,$$

where the dot denotes the scalar product in R^M. We also recall that a symmetric matrix S is positive definite if and only if the eigenvalues of S are strictly positive.

Since a positive definite matrix is non-singular it follows that the linear system (1.5) has a unique solution. We also note that A is *sparse,* ie, only a few elements of A are different from zero (A is tridigagonal). This very important property depends, as we have seen, on the fact that a basis function φ_j of V_h is different from zero only on a few intervals and thus will interfere only with a few other basis functions. The fact that the basis functions may be chosen in this way is an important distinctive feature of the finite element method.

In the special case of a uniform partition with $h_j = h = \frac{1}{M+1}$ the system (1.5) takes the form

$$(1.6) \qquad \frac{1}{h}\begin{bmatrix} 2 & -1 & 0 & \cdots & \cdots & 0 \\ -1 & 2 & -1 & 0 & & \vdots \\ 0 & -1 & 2 & \ddots & \ddots & \vdots \\ \vdots & 0 & \ddots & \ddots & \ddots & 0 \\ \vdots & & \ddots & \ddots & 2 & -1 \\ 0 & \cdots & \cdots & 0 & -1 & 2 \end{bmatrix} \begin{bmatrix} \xi_1 \\ \\ \\ \\ \\ \xi_M \end{bmatrix} = \begin{bmatrix} b_1 \\ \\ \\ \\ \\ b_M \end{bmatrix}.$$

After division by h this may be interpreted as a variant of a standard difference method for (D) where the elements of the right hand side b_j/h are mean values of f over the intervals (x_{j-1}, x_{j+1}) (cf Problem 1.4).

To sum up, we have seen that the finite element method (V_h) for (D) leads to a linear system of equations with a sparse, symmetric and positive definite stiffness matrix.

Problems

1.3 Construct a finite-dimensional subspace V_h of V consisting of functions which are quadratic on each subinterval I_j of a partition of $I=(0, 1)$. How can one choose the parameters to describe such functions? Find the corresponding basis functions. Then formulate a finite element method for (D) using the space V_h and write down the corresponding linear system of equations in case of a uniform partition.

1.4 Formulate a difference method for (D) and compare with (1.6).

1.5 Consider the boundary value problem

$$\frac{d^4u}{dx^4}=f, \qquad 0<x<1, \tag{1.7}$$
$$u(0)=u'(0)=u(1)=u'(1)=0.$$

Here u represents e g the deflection of a clamped beam subject to a transversal force with intensity f (see Fig 1.5).

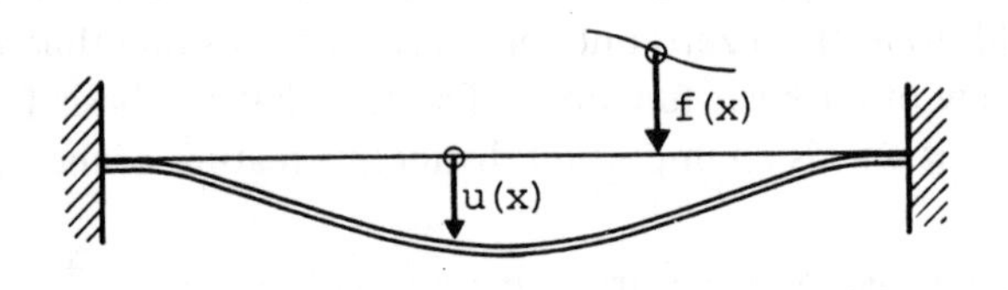

Fig 1.5

(a) In mechanics this beam problem would naturally be formulated as follows:

$$M=u'', \qquad 0<x<1, \tag{1.8a}$$
$$M''=f, \qquad 0<x<1, \tag{1.8b}$$
$$u(0)=u'(0)=u(1)=u'(1)=0. \tag{1.8c}$$

What does here the quanity M represent and what is the mechanical interpretation of (1.8a–c)?

(b) Show that the problem (1.7) can be given the following variational formulation: Find $u \in W$ such that

$$(u'', v'') = (f, v) \qquad \forall v \in W,$$

where $W = \{v$: v and v' are continuous on $[0,1]$, v'' is piecewise continuous and $v(0) = v'(0) = v(1) = v'(1) = 0\}$.

(c) For $I = [a, b]$ an interval, define

$$P_3(I) = \{v\text{: v is a polynomial of degree} \leq 3 \text{ on I, i e, v has the form } v(x) = a_3x^3 + a_2x^2 + a_1x + a_0,\ x \in I \text{ where } a_i \in R\}.$$

Show that $v \in P_3(I)$ is uniquely determined by the values $v(a)$, $v'(a)$, $v(b)$, $v'(b)$. Find the corresponding basis functions (the basis function corresponding to the value $v(a)$ is the cubic polynomial v such that $v(a) = 1$, $v'(a) = 0$, $v(b) = v'(b) = 0$, etc).

(d) Starting from (c) construct a finite-dimensional subspace W_h of W consisting of piecewise cubic functions. Specify suitable parameters to describe the functions in W_h and determine the corresponding basis functions.

(e) Formulate a finite element method for (1.7) based on the space W_h. Find the corresponding linear system of equations in the case of a uniform partition. Determine the solution in e g the case of two intervals and f constant. Compare with the exact solution.

1.3 An error estimate for FEM for the model problem

We shall now study the error $u - u_h$ where u is the solution of (D) and u_h is the solution of the finite element problem (V_h), i e, $u_h \in V_h$ and u_h satisfies (1.2). The proof is based on the following equation for the error:

(1.9) $$((u - u_h)', v') = 0 \qquad \forall v \in V_h.$$

This follows by recalling that $(u', v') = (f, v)$, $\forall v \in V$, so that in particular since $V_h \subset V$

(1.10) $$(u', v') = (f, v) \qquad \forall v \in V_h.$$

Subtracting (1.2) from (1.10) we obtain (1.9).

We shall use the notation

$$||w||=(w, w)^{1/2}=(\int_0^1 w^2dx)^{1/2};$$

$||\cdot||$ is the *norm* associated with the *scalar product* (.,.). We also recall *Cauchy's inequality:*

$$(1.11) \qquad |(v, w)|\leq||v|| \, ||w||.$$

We shall prove the following estimate for $u-u_h$ which shows that in a certain sense u_h is the best possible approximation to the exact solution u.

Theorem 1.1. For any $v\in V_h$ we have

$$||(u-u_h)'||\leq||(u-v)'||.$$

Proof Let $v\in V_h$ be arbitrary and set $w=u_h-v$. Then $w\in V_h$ and using (1.9) with v replaced by w, we get, using Cauchy's inequality also,

$$\begin{aligned}||(u-u_h)'||^2&=((u-u_h)', (u-u_h)')+((u-u_h)', w')\\&=((u-u_h)', (u-u_h+w)')=((u-u_h)', (u-v)')\\&\leq||(u-u_h)'|| \, ||(u-v)'||.\end{aligned}$$

Dividing by $||(u-u_h)'||$ we obtain the statement of the theorem (if $||(u-u_h)'||=0$, then the theorem clearly holds). □

From Theorem 1.1 we can obtain a quantitative estimate for the error $||(u-u_h)'||$ by estimating $||(u-\tilde{u}_h)'||$ where $\tilde{u}_h\in V_h$ is a suitably chosen function. We shall choose $\tilde{u}_h\in V_h$ to be the *interpolant* of u, i e, $\tilde{u}_h$ interpolates u at the nodes x_j, i e,

$$\tilde{u}_h(x_j)=u(x_j) \qquad j=0,\ldots, M+1.$$

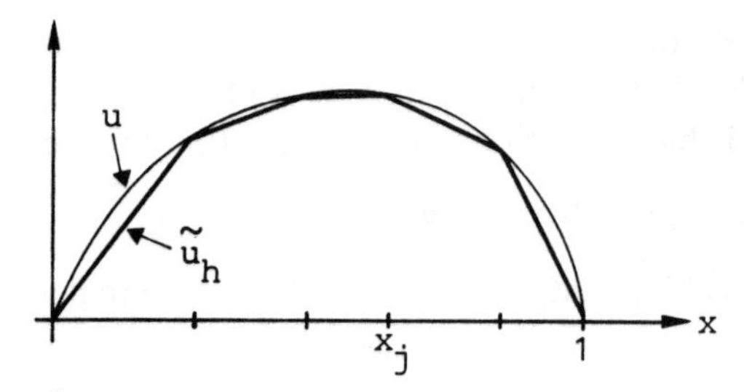

Fig 1.6 The interpolant $\tilde{u}_h$.

It is easy to see (cf any basic course in numerical analysis or Problem 4.1 below) that if $\tilde{u}_h \in V_h$ is chosen in this way, then for $0 \le x \le 1$,

$$(1.12) \qquad |u'(x) - \tilde{u}_h'(x)| \le h \max_{0 \le y \le 1} |u''(y)|,$$

$$(1.13) \qquad |u(x) - \tilde{u}_h(x)| \le \frac{h^2}{8} \max_{0 \le y \le 1} |u''(y)|.$$

Using (1.12) and Theorem 1.1 we now obtain the following estimate for the derivative of the error $u - u_h$:

$$(1.14) \qquad \|(u - u_h)'\| \le h \max_{0 \le y \le 1} |u''(y)|.$$

Since $(u - u_h)(0) = 0$ we obtain from (1.14) by integration the following estimate for the error itself (cf Problem 1.6):

$$(1.15) \qquad |u(x) - u_h(x)| \le h \max_{0 \le y \le 1} |u''(y)| \quad \text{for } 0 \le x \le 1.$$

We observe that this latter estimate is less sharp than the estimate (1.13) for the interpolation error where we have a factor h^2. With a more precise analysis it is possible to show that in fact also the finite element method gives a factor h^2 for the error $u - u_h$ (cf also Problem 1.19 below).

Let us note that the quantity u', representing a deformation or a force in Examples A and B above, is usually of more (or at least no less) practical interest than the quantity u itself, representing in these cases a displacement. Thus the estimate (1.14) is of independent interest and not just a step on the way to an estimate of $u - u_h$.

Let us also notice that to prove (1.14) we do not need to concretely construct $\tilde{u}_h$ (which would require knowledge of the exact solution u); we only have to be able to give an estimate of the interpolation error, for instance of the form (1.12), (1.13).

To sum up, by Theorem 1.1 we have the qualitative information that $\|(u - u_h)'\|$ is "as small as possible" and by using also the interpolation estimate (1.12) we obtain the quantitative error estimate (1.14), which in particular shows that the error tends to zero as the maximum length of the subintervals I_j tends to zero if u'' is bounded on $[0,1]$.

Problem

1.6 Prove (1.15) using (1.14) and the boundary conditions $u(0) = u_h(0) = 0$. Hint: Use the relation

$$(u - u_h)(x) = \int_0^x (u - u_h)'(y)\,dy$$

together with Cauchy's inequality.

1.4 FEM for the Poisson equation

We will now consider the following boundary value problem for the Poisson equation:

$$(1.16a) \qquad -\Delta u = f \quad \text{in } \Omega,$$

$$(1.16b) \qquad u = 0 \quad \text{on } \Gamma,$$

where Ω is a bounded open domain in the plane $R^2 = \{x = (x_1, x_2): x_i \in R\}$ with boundary Γ, f is a given function and as usual,

$$\Delta u = \frac{\partial^2 u}{\partial x_1^2} + \frac{\partial^2 u}{\partial x_2^2}.$$

A number of problems in physics and mechanics are modelled by (1.16); u may represent for instance a temperature, an electro-magnetic potential or the displacement of an elastic membrane fixed at the boundary under a transversal load of intensity f (see Fig 1.7 and compare also with problem B of Section 1.1).

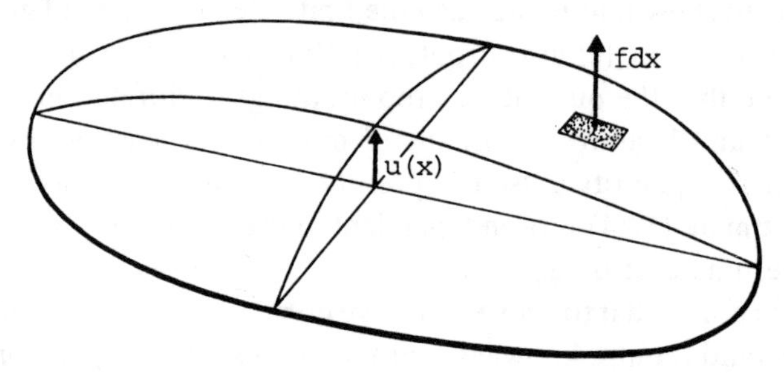

Fig 1.7

Let us now before continuing recall a certain *Green's formula* which will be of fundamental importance in what follows. Let us start from the *divergence theorem* (in two dimensions):

$$\int_\Omega \text{div } A \, dx = \int_\Gamma A \cdot n \, ds,$$

where $A = (A_1, A_2)$ is a vector-valued function defined on Ω,

$$\text{div } A = \frac{\partial A_1}{\partial x_1} + \frac{\partial A_2}{\partial x_2},$$

and $n = (n_1, n_2)$ is the outward unit normal to Γ. Here dx denotes the element of area in R^2 and ds the element of arc length along Γ. If we apply the divergence theorem to $A = (vw, 0)$ and $A = (0, vw)$, we find that

$$(1.17)\qquad \int_\Omega \frac{\partial v}{\partial x_i} w\,dx + \int_\Omega v \frac{\partial w}{\partial x_i}\,dx = \int_\Gamma v w n_i\,ds, \qquad i=1,2.$$

Denoting by ∇v the *gradient* of v, ie, $\nabla v = \left(\frac{\partial v}{\partial x_1}, \frac{\partial v}{\partial x_2}\right)$, we get from (1.17) the following Green's formula:

$$\int_\Omega \nabla v \cdot \nabla w\,dx \equiv \int_\Omega \left[\frac{\partial v}{\partial x_1}\frac{\partial w}{\partial x_1} + \frac{\partial v}{\partial x_2}\frac{\partial w}{\partial x_2}\right] dx$$

$$= \int_\Gamma \left[v\frac{\partial w}{\partial x_1} n_1 + v\frac{\partial w}{\partial x_2} n_2\right] ds - \int_\Omega v\left[\frac{\partial^2 w}{\partial x_1^2} + \frac{\partial^2 w}{\partial x_2^2}\right] dx$$

$$= \int_\Gamma v\frac{\partial w}{\partial n}\,ds - \int_\Omega v \Delta w\,dx,$$

ie,

$$(1.18)\qquad \int_\Omega \nabla v \cdot \nabla w\,dx = \int_\Gamma v\frac{\partial w}{\partial n}\,ds - \int_\Omega v\Delta w\,dx,$$

where

$$\frac{\partial w}{\partial n} = \frac{\partial w}{\partial x_1} n_1 + \frac{\partial w}{\partial x_2} n_2$$

is the *normal derivative,* ie, the derivative in the outward normal direction to the boundary Γ.

We shall now give a variational formulation of problem (1.16). We shall first show that if u satisfies (1.16), then u is the solution of the following variational problem: Find $u \in V$ such that

$$(1.19)\qquad a(u, v) = (f, v) \qquad \forall v \in V,$$

where

$$a(u, v) = \int_\Omega \nabla u \cdot \nabla v\,dx = \int_\Omega \left[\frac{\partial u}{\partial x_1}\frac{\partial v}{\partial x_1} + \frac{\partial u}{\partial x_2}\frac{\partial v}{\partial x_2}\right] dx,$$

$$(f, v) = \int_\Omega f v\,dx,$$

$$V = \{v: v \text{ is continuous on } \Omega, \frac{\partial v}{\partial x_1} \text{ and } \frac{\partial v}{\partial x_2} \text{ are piecewise continuous on } \Omega \text{ and } v=0 \text{ on } \Gamma\}.$$

In exactly the same way as in Section 1.1, we see that $u \in V$ satisfies (1.19) if and only if u is the solution of the following minimization problem: Find $u \in V$ such that $F(u) \leq F(v)$, $\forall v \in V$, where $F(v)$ is the total potential energy

$$F(v) = \frac{1}{2} a(v, v) - (f, v).$$

To see that (1.19) follows from (1.16) we multiply (1.16a) with an arbitrary test function $v \in V$ and integrate over Ω. According to Green's formula (1.18) we then have

$$(f, v) = -\int_{\Omega} \Delta u \, v dx = -\int_{\Gamma} \frac{\partial u}{\partial n} v \, ds + \int_{\Omega} \nabla u \cdot \nabla v \, dx = a(u, v),$$

where the boundary integral vanishes since $v=0$ on Γ. On the other hand, if $u \in V$ satisfies (1.19) and u is sufficiently regular, then we see as in Section 1.1 that u also satisfies (1.16) (cf Problem 1.10).

Let us now construct a finite-dimensional subspace V_h of V. For simplicity we shall assume that Γ is a polygonal curve, in which case we say that Ω is a polygonal domain (if Γ is curved we may first approximate Γ with a polygonal curve, see Chapter 12). Let us now make a *triangulation* of Ω, by subdividing Ω into a set $T_h = K_1, \ldots, K_m$ of non-overlapping triangles K_i,

$$\Omega = \bigcup_{K \in T_h} K = K_1 \cup K_2 \ldots \cup K_m,$$

such that no vertex of one triangle lies on the edge of another triangle (see Fig 1.8)

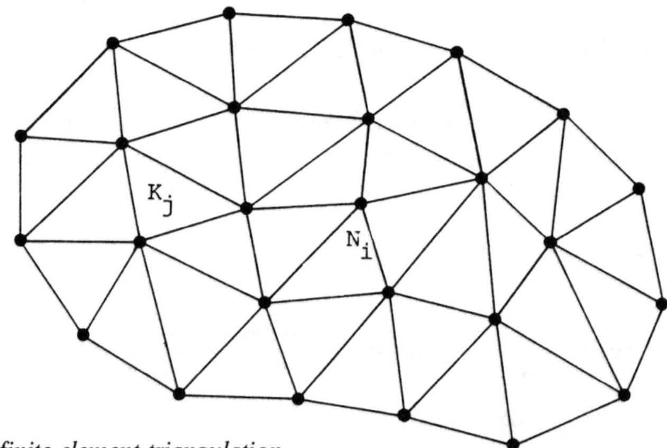

Fig 1.8 A finite element triangulation

We introduce the mesh parameter

$$h = \max_{K \in T_h} \text{diam}(K), \ \text{diam}(K) = \text{diameter of } K = \text{longest side of } K.$$

We now define V_h as follows:

$$V_h = \{v: v \text{ is continuous on } \Omega, \ v|_K \text{ is linear for } K \in T_h, \ v=0 \text{ on } \Gamma\}.$$

Here $v|_K$ denotes the restriction of v to K, ie, the function defined on K agreeing with v on K. The space V_h consists of all continuous functions that are linear on each triangle K and vanish on Γ. We notice that $V_h \subset V$. As parameters to describe a function $v \in V_h$ we choose the values $v(N_i)$ of v at the *nodes* N_i, $i=1, \ldots, M$, of T_h (see Fig 1.8) but exclude the nodes on the boundary since $v=0$ on Γ. The corresponding basis functions $\varphi_j \in V_h$, $j=1, \ldots, M$, are then defined by (see Fig 1.9)

$$\varphi_j(N_i)=\delta_{ij} \equiv \begin{cases} 1 & \text{if } i=j \\ 0 & \text{if } i \neq j \end{cases} \qquad i, j=1, \ldots, M.$$

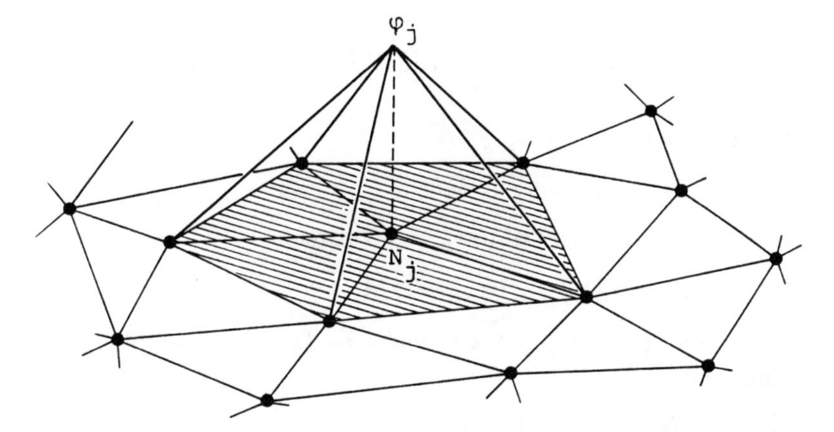

Fig 1.9 The basis function φ_j.

We see that the *support* of φ_j (the set of points x for which $\varphi_j(x) \neq 0$) consists of the triangles with the common node N_j (the shaded area in Fig 1.9). A function $v \in V_h$ now has the representation

$$v(x)=\sum_{j=1}^{M} \eta_j \varphi_j(x), \quad \eta_j=v(N_j), \text{ for } x \in \Omega \cup \Gamma.$$

We can now formulate the following finite element method for (1.16) starting from the variational formulation (1.19): Find $u_h \in V_h$ such that

$$(1.20) \qquad a(u_h, v)=(f, v) \quad \forall v \in V_h.$$

Exactly as in Section 1.2 we see that (1.20) is equivalent to the linear system of equations

$$(1.21) \qquad A\xi=b,$$

where $A=(a_{ij})$, the stiffness matrix, is an $M\times M$ matrix with elements $a_{ij}=a(\varphi_i, \varphi_j)$ and $\xi=(\xi_i)$, $b=(b_i)$ are M-vectors with elements $\xi_i=u_h(N_i)$, $b_i=(f, \varphi_i)$.

Clearly A is symmetric and as in Section 1.2 we see that A is positive definite and thus in particular non-singular so that (1.21) admits a unique solution ξ. Moreover, A is again sparse; we have that $a_{ij}=0$ unless N_i and N_j are nodes of the same triangle.

In the same way as in Section 1.2 we realize that $u_h\in V_h$ is the best approximation of the exact solution u in the sense that

$$(1.22) \qquad ||\nabla u-\nabla u_h||\leqslant||\nabla u-\nabla v|| \qquad \forall v\in V_h,$$

where

$$||\nabla v||=a(v,v)^{1/2}=(\int_\Omega|\nabla v|^2dx)^{1/2}.$$

In particular we have

$$(1.23) \qquad ||\nabla u-\nabla u_h||\leqslant||\nabla u-\nabla \tilde{u}_h||,$$

where $\tilde{u}_h$ is the interpolant of u, i e, $\tilde{u}_h\in V_h$ and

$$\tilde{u}_h(N_i)=u(N_i) \qquad i=1, \ldots, M.$$

In Chapter 4 we prove that if the triangles $K\in T_h$ are not allowed to become too thin, then

$$(1.24) \qquad ||\nabla u-\nabla \tilde{u}_h||\leqslant Ch.$$

Here and below we denote by C a positive constant, possibly different at different occurences, that does not depend on the mesh parameter h. In the case (1.24) the constant C depends on the size of the second partial derivatives of u and the smallest angle of the triangles $K\in T_h$. One can also prove (see Section 4.7) that

$$||u-u_h||\equiv(\int_\Omega(u-u_h)^2dx)^{1/2}\leqslant Ch^2$$

with a similar dependence of C. In particular these estimates show that if the exact solution u is sufficiently regular, then the error and the gradient of the error $u-u_h$ tend to zero in the norm $||\cdot||$ as h tends to zero.

Example 1.1. Let Ω be a square with side length 1 and let T_h be the uniform triangulation of Ω according to Fig 1.10 with the indicated enumeration of the nodes of T_h.

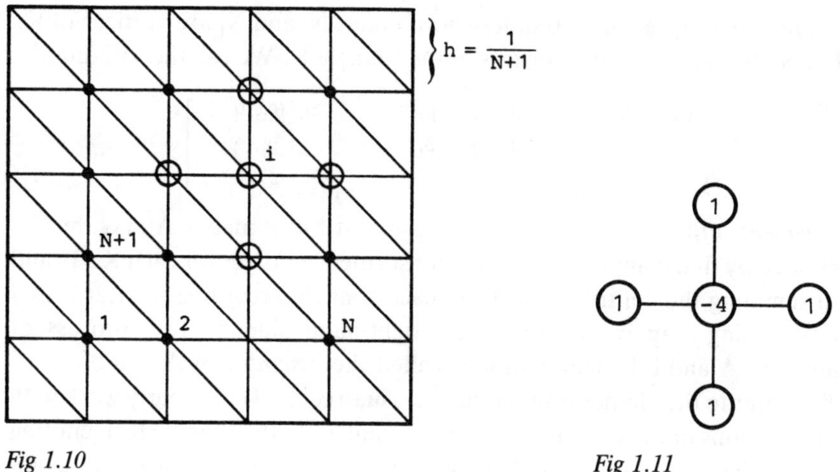

Fig 1.10

Fig 1.11

In this case the linear system (1.21) reads as follows:

$$
(1.25) \qquad \text{row } (N+1) \begin{bmatrix} 4 & -1 & 0 & -1 & 0 & \cdot & \cdot & 0 \\ -1 & 4 & -1 & 0 & -1 & 0 & \cdot & \cdot \\ 0 & -1 & 4 & -1 & 0 & -1 & \cdot & \cdot \\ -1 & 0 & -1 & 4 & -1 & 0 & -1 & 0 \\ 0 & \cdot & \cdot & \cdot & \cdot & \cdot & 0 & -1 \\ \cdot & 0 & \cdot & \cdot & \cdot & \cdot & -1 & 0 \\ \cdot & \cdot & \cdot & -1 & 0 & -1 & 4 & -1 \\ 0 & \cdot & \cdot & 0 & -1 & 0 & -1 & 4 \end{bmatrix} \begin{bmatrix} \xi_1 \\ \\ \cdot \\ \cdot \\ \cdot \\ \\ \\ \xi_M \end{bmatrix} = \begin{bmatrix} b_1 \\ \\ \cdot \\ \cdot \\ \cdot \\ \\ \\ b_M \end{bmatrix}.
$$

Note that here the left-hand side of equation i is a linear combination of the values of u_h at the 5 nodes indicated in Fig 1.10 with coefficients given in Fig 1.11. Dividing by h^2 we recognize this as the linear system obtained by applying the so-called 5-point difference method for (1.16) with the components of the right-hand side being weighted averages of f around the nodes N_i (cf Problem 1.7 below). □

The elements $a_{ij}=a(\varphi_i, \varphi_j)$ in the stiffness matrix A are usually in practice computed by summing the contributions from the different triangles:

$$
(1.26) \qquad a(\varphi_i, \varphi_j)=\sum_{K\in T_h} a_K(\varphi_i, \varphi_j),
$$

where

$$
a_K(\varphi_i, \varphi_j)=\int_K \nabla\varphi_i \cdot \nabla\varphi_j dx.
$$

We notice that $a_K(\varphi_i, \varphi_j)=0$ unless both nodes N_i and N_j are vertices of K.

Let N_i, N_j and N_k be the vertices of the triangle K. We call the 3×3-matrix

$$\begin{bmatrix} a_K(\varphi_i, \varphi_i) & a_K(\varphi_i, \varphi_j) & a_K(\varphi_i, \varphi_k) \\ & a_K(\varphi_j, \varphi_j) & a_K(\varphi_j, \varphi_k) \\ \text{sym} & & a_K(\varphi_k, \varphi_k) \end{bmatrix} \tag{1.27}$$

the *element stiffness matrix* for K. The global stiffness matrix A may thus be computed by first computing the element stiffness matrices for each $K \in T_h$ and then summing the contributions from each triangle according to (1.26). In a corresponding way we compute the right-hand side b. This process of computing A and b by summation is called the *assembly* of A and b.

To compute the elements in the stiffness matrix (1.27) we clearly work with the restrictions of the basis functions φ_i, φ_j and φ_k to the triangle K. Denoting these restrictions by ψ_i, ψ_j and ψ_k, we have that each ψ is a linear function on K that takes the value one at one vertex and vanishes at the other two vertices of K. We call ψ_i, ψ_j and ψ_k the *basis functions on the triangle* K, cf Fig 1.12. If w is a linear function on K, then w has the representation

$$w(x)=w(N_i)\psi_i(x)+w(N_j)\psi_j(x)+w(N_k)\psi_k(x), \qquad x \in K.$$

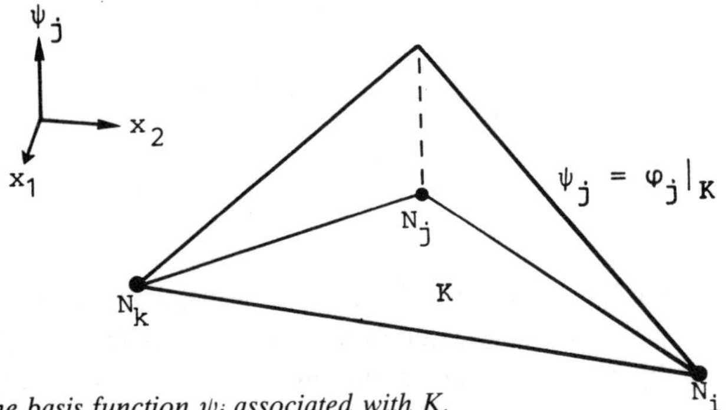

Fig 1.12 The basis function ψ_j associated with K.

Problems

1.7 Formulate a difference method for (1.16) in the case when Ω is a square using the difference approximation

$$\frac{\partial^2 u}{\partial x_1^2}(x_1, x_2) \sim \frac{u(x_1+h, x_2)-2u(x_1, x_2)+u(x_1-h, x_2)}{h^2},$$

and a corresponding approximation for $\frac{\partial^2 u}{\partial x_2^2}$. Compare with Example 1.1.

1.8 Find the linear basis functions for the triangle K with vertices at (0, 0), (h, 0) and (0, h). Show that the corresponding element stiffness matrix (1.27) is given by

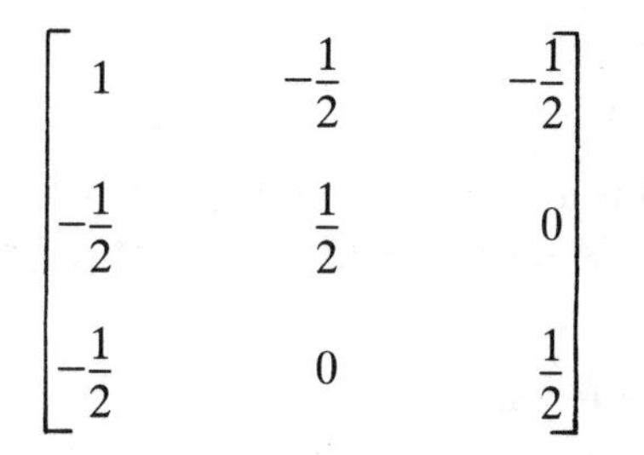

$$\begin{bmatrix} 1 & -\frac{1}{2} & -\frac{1}{2} \\ -\frac{1}{2} & \frac{1}{2} & 0 \\ -\frac{1}{2} & 0 & \frac{1}{2} \end{bmatrix}$$

Using this result show that the linear system (1.25) of Example 1.1 has the stated form.

1.9 Find the element stiffness matrix (1.27) for a general triangle K in terms of the coordinates $a^i=(a_1^i, a_2^i)$, $i=1, 2, 3$, of the vertices of K.

1.10 Show that if $u \in V$ satisfies (1.19) and u is twice continuously differentiable, then u satisfies (1.16).

1.11 Find the element stiffness matrix for the problem

$$-u''=f \qquad \text{for } 0<x<1, \quad u(0)=u(1)=0,$$

if we use piecewise quadratic functions according to Problem 1.3. Then determine the corresponding global stiffness matrix in the case of a uniform subdivision. Can you interpret the resulting equations as difference approximations of the equation $-u''=f$?

1.5 The Hilbert spaces $L_2(\Omega)$, $H^1(\Omega)$ and $H_0^1(\Omega)$

When giving variational formulations of boundary value problems for partial differential equations, it is from the mathematical point of view natural and very useful to work with function spaces V that are slightly larger (i e contain somewhat more functions) than the spaces of continuous functions with piecewise continuous derivatives used in the preceeding sections. It is also useful to endow the spaces V with various *scalar products* with the scalar product related to the boundary value problem. More precisely, V will be a *Hilbert space,* (see below).

Before introducing these Hilbert spaces let us recall a few simple concepts from linear algebra: If V is a linear space, then we say that L is a *linear form* on V if L: $V \to R$, i e, $L(v) \in R$ for $v \in V$, and L is *linear,* i e, for all v, $w \in V$ and $\beta, \theta \in R$

$$L(\beta v+\theta w)=\beta L(v)+\theta L(w).$$

Furthermore, we say that a(. , .) is a *bilinear form* on V×V if a: $V \times V \to R$, i e, $a(v, w) \in R$ for v, $w \in V$, and a is linear in each argument, i e, for all u, v, $w \in V$ and $\beta, \theta \in R$ we have

$$a(u, \beta v+\theta w)=\beta a(u, v)+\theta a(u, w),$$
$$a(\beta u+\theta v, w)=\beta a(u, w)+\theta a(v, w).$$

The bilinear form a(. , .) on V×V is said to be *symmetric* if

$$a(v,w)=a(w,v) \qquad \forall v, w \in V.$$

A symmetric bilinear form a(. , .) on V×V is said to be a *scalar product* on V if

$$a(v, v)>0 \qquad \forall v \in V, v \neq 0.$$

The *norm* $\|\cdot\|_a$ associated with a scalar product a(. , .) is defined by

$$\|v\|_a=(a(v, v))^{1/2}, \qquad \forall v \in V.$$

Further, if $<. , .>$ is a scalar product with corresponding norm $\|\cdot\|$, then we have *Cauchy's inequality*

$$(1.28) \qquad |<v, w>| \leqslant \|v\| \, \|w\|.$$

We further recall that if V is a linear space with a scalar product with corresponding norm $\|\cdot\|$, then V is said to be a *Hilbert space* if V is complete, i e, if every *Cauchy sequence* with respect to $\|\cdot\|$ is convergent. We recall that a sequence $v_1, v_2, v_3, \ldots$, of elements v_i in the space V with norm $\|\cdot\|$ is said to be a Cauchy sequence if for all $\varepsilon>0$ there is a natural number N such that $\|v_i-v_j\|<\varepsilon$ if $i, j>N$. Further, v_i converges to v if $\|v-v_i\| \to 0$ as $i \to \infty$. The reader unfamiliar with the concept of completeness may bypass this remark and think of a Hilbert space simply as a linear space with a scalar product.

We now introduce some Hilbert spaces that are natural to use for variational formulations of the boundary value problems we will consider. Let us start with the one-dimensional case. If $I=(a, b)$ is an interval, we define the space of "square integrable functions" on I:

$$L_2(I)=\{v: v \text{ is defined on I and } \int_I v^2 dx<\infty\}.$$

The space $L_2(I)$ is a Hilbert space with the scalar product

$$(v, w)=\int_I vw\, dx,$$

and the corresponding norm (the L_2-norm):

$$\|v\|_{L_2(I)}=\left(\int_I v^2dx\right)^{1/2}=(v, v)^{1/2}.$$

By Cauchy's inequality,

$$|(v, w)|\leqslant\|v\|_{L_2(I)}\|w\|_{L_2(I)},$$

we see that (v, w) is well-defined, i e, the integral (v, w) exists, if v and $w\in L_2(I)$.

Remark. To really appreciate the definition of $L_2(I)$ and realize that this space is complete requires some familiarity with the Lebesgue integral. In this book, however, it is sufficient to get an idea of $L_2(I)$ by using the usual Riemann integral; from this point of view we may think of a "typical" function $v\in L_2(I)$ as a piecewise continuous function, possibly unbounded, such that $\int_I v^2dx<\infty$. □

Example 1.2 We have that the function $v(x)=x^{-\beta}$, $x\in I=(0, 1)$ belongs to $L_2(I)$ if $\beta<\frac{1}{2}$. □

We also introduce the space $H^1(I)=\{v: v \text{ and } v' \text{ belong to } L_2(I)\}$, and we equip this space with the scalar product

$$(v,w)_{H^1(I)}=\int_I (vw+v'w')dx,$$

and the corresponding norm

$$\|v\|_{H^1(I)}=\left(\int_I [v^2+(v')^2]dx\right)^{1/2}.$$

The space $H^1(I)$ thus consists of the functions v defined on I which together with their first derivatives are square-integrable, i e, belong to $L_2(I)$.

In the case of boundary value problems of the form $-u''=f$ on $I=(a, b)$ with boundary conditions $u(a)=u(b)=0$, we shall use the space

$$H_0^1(I)=\{v\in H^1(I): v(a)=v(b)=0\}$$

with the same scalar product and norm as for $H^1(I)$.

Our introductory boundary value problem

$$(1.29)\qquad \begin{aligned} -u''&=f \quad \text{on } I=(0,1),\\ u(0)&=u(1)=0, \end{aligned}$$

can now be given the following variational formulation:

$$(1.30)\qquad \text{Find } u\in H_0^1(I) \text{ such that } (u', v')=(f, v) \qquad \forall v\in H_0^1(I),$$

with (.,.) as in Section 1.1. If we compare (1.30) with the formulation (*V*) in Section 1.1, we note that the space $H_0^1(I)$ is larger than the space V used in the formulation (*V*). The space $H_0^1(I)$ is specially tailored for a variational formulation of (1.29) and is in fact the largest space for which a variational formulation of the form (1.30) is meaningful. From a mathematical point of view the "right" choice of function space is essential since this may make it easier to prove the existence of a solution to the continuous problem. From the finite element point of view the formulation (1.30) as opposed to (*V*) is of interest mainly because the basic error estimate for the finite element method is an estimate in the norm indicated by (1.30) (the $H^1(I)$-norm). Further, using the standard notation $L_2(I)$, $H^1(I)$, $H_0^1(I)$ etc, we may give our boundary value problems variational formulations in a concise way, as will be seen below.

Now let Ω be a bounded domain R^d, d=2 or 3, and define

$$L_2(\Omega)=\{v\colon v \text{ is defined on } \Omega \text{ and } \int_\Omega v^2dx<\infty\},$$

$$H^1(\Omega)=\{v\in L_2(\Omega)\colon \frac{\partial v}{\partial x_i}\in L_2(\Omega),\ i=1,\ldots, d\},$$

and introduce the corresponding scalar products and norms

$$(v, w)=\int_\Omega vw\,dx,\quad \|v\|_{L_2(\Omega)}=(\int_\Omega v^2dx)^{1/2},$$

$$(v, w)_{H^1(\Omega)}=\int_\Omega[vw+\nabla v\cdot\nabla w]dx,$$

$$\|v\|_{H^1(\Omega)}=(\int_\Omega[v+|\nabla v|^2]dx)^{1/2}.$$

We also define

$$H_0^1(\Omega)=\{v\in H^1(\Omega)\colon v=0 \text{ on } \Gamma\},$$

where Γ is the boundary of Ω and we equip $H_0^1(\Omega)$ with the same scalar product and norm as $H^1(\Omega)$.

The boundary value problem

$$(D) \qquad \begin{aligned} -\Delta u &= f \quad \text{in } \Omega, \\ u &= 0 \quad \text{on } \Gamma, \end{aligned}$$

can now be given the following variational formulation:

(V) Find $u \in H_0^1(\Omega)$ such that $a(u,v)=(f,v) \quad \forall v \in H_0^1(\Omega)$,

or equivalently

(M) Find $u \in H_0^1(\Omega)$ such that $F(u) \leq F(v) \quad \forall v \in H_0^1(\Omega)$,

where

$$F(v)=\frac{1}{2}a(v,v)-(f,v),$$

$$a(u,\, v)=\int_\Omega \nabla u \cdot \nabla v dx, \quad (f,v)=\int_\Omega fv\, dx.$$

Remark The formulation (V) is said to be a *weak formulation* of (D) and the solution of (V) is said to be a *weak solution* of (D). If u is a weak solution of (D) then it is not immediately clear that u is also a classical solution of (D), since this requires u to be sufficiently regular so that Δu is defined in a classical sense. The advantage mathematically of the weak formulation (V) is that it is easy to prove the existence of a solution to (V), whereas it is relatively difficult to prove the existence of a classical solution to (D). To prove the existence of a classical solution of (D) one usually starts with the weak solution of (D) and shows, often with considerable effort, that in fact this solution is sufficiently regular to be also a classical solution. For more complicated, e g non-linear problems, it may be extremely difficult or practically impossible to prove the existence of classical solutions whereas existence of weak solutions may still be within reach. □

Problems

1.12 Let $\Omega=\{x \in R^2 : |x| \leq 1\}$. Show that the function $v(x)=|x|^\alpha$ belongs to $H^1(\Omega)$ if $\alpha>0$.

1.13 Prove Cauchy's inequality (1.28).

1.14 Consider the problem corresponding to (D) with an inhomogeneous boundary condition, i e, the problem

$$(1.31) \qquad \begin{aligned} -\Delta u &= f \quad \text{in } \Omega, \\ u &= u_0 \quad \text{on } \Gamma, \end{aligned}$$

where f and u_0 are given. Show that this problem can be given the following equivalent variational formulations:

(V) Find $u \in V(u_0)$ such that $a(u,v)=(f,v) \quad \forall v \in H_0^1(\Omega)$,
(M) Find $u \in V(u_0)$ such that $F(u) \leqslant F(v) \quad \forall v \in V(u_0)$,

where

$$V(u_0)=\{v \in H^1(\Omega): v=u_0 \text{ on } \Gamma\}.$$

Then formulate a finite element method for (1.31) and prove an error estimate.

1.6 A geometric interpretation of FEM

We shall now give an interpretation of the finite element method in geometric terms in the function space $H_0^1(\Omega)$. We recall that two elements v and w in a linear space with scalar product $<.,.>$ are said to be *orthogonal* if $<v, w>=0$.

Let us for simplicity consider the following variant of our previous problem (1.16):

$$\begin{aligned} -\Delta u+u&=f \quad \text{in } \Omega, \\ u&=0 \quad \text{on } \Gamma, \end{aligned} \tag{1.32}$$

(cf Problem 2.5 below). The corresponding variational problem reads: Find $u \in H_0^1(\Omega)$ such that

$$\int_\Omega \nabla u \cdot \nabla v dx+\int_\Omega uv\, dx=(f, v) \qquad \forall v \in H_0^1(\Omega),$$

or

$$<u, v>=(f, v) \qquad \forall v \in H_0^1(\Omega), \tag{1.33}$$

using the notation

$$<u, v>=\int_\Omega [\nabla u \cdot \nabla v+uv]dx.$$

Note that $<.,.>$ is in fact the scalar product in the space $H_0^1(\Omega)$.

Let V_h be a finite-dimensional subspace of $H_0^1(\Omega)$, e g the space of piecewise linear functions of Section 1.4, and consider the following finite element method for (1.32): Find $u_h \in V_h$ such that

$$<u_h, v>=(f, v) \quad \forall v \in V_h. \tag{1.34}$$

Since $V_h \subset H_0^1(\Omega)$ we may choose $v \in V_h$ in (1.33) and on substraction from (1.34), we obtain

$$(1.35) \qquad <u-u_h, v>=0 \quad \forall v \in V_h,$$

ie the error $u-u_h$ is orthogonal to V_h with respect to $<.,.>$. We may also express this fact as follows: The finite element solution u_h is the *projection* with respect to $<.,.>$ of the exact solution u *on* V_h, ie, u_h is the element in V_h closest to u with respect to the $H^1(\Omega)$-norm $\|\cdot\|_{H^1(\Omega)}$, or in other words

$$(1.36) \qquad \|u-u_h\|_{H^1(\Omega)} \leq \|u-v\|_{H^1(\Omega)} \quad \forall v \in V_h.$$

This situation is symbolically illustrated in Fig 1.13 where $H_0^1(\Omega)$ is represented by the whole plane while the straight line through the origin represents V_h.

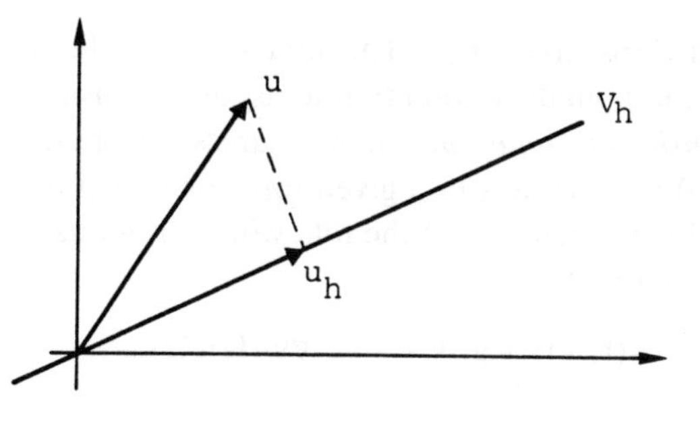

Fig 1.13

According to (1.36), u_h is the best approximation of the exact solution u, in the sense that for no other function $v \in V_h$, is the error $u-v$ smaller when measured in the $H^1(\Omega)$-norm. We have seen that u_h can be found by solving a linear system of equations with right hand side depending on the given function f. Thus, we can compute a best approximation u_h of u, without knowing u itself, knowing only that $-\Delta u+u=f$ in Ω and $u=0$ on Γ. This remarkable fact reflects the ellipticity of the boundary value problem (1.32).

Problem

1.15 Prove that (1.35) and (1.36) are equivalent (cf the proof of Theorem 1.1).

1.7 A Neumann problem. Natural and essential boundary conditions

We shall now consider a problem with another type of boundary condition, namely the following *Neumann problem* (*D*):

(1.37a) $\quad -\Delta u+u=f \qquad$ in Ω

(1.37b) $\quad \frac{\partial u}{\partial n}=g \qquad$ on Γ,

where again Ω is a bounded domain with boundary Γ and $\frac{\partial}{\partial n}$ denotes the outward normal derivative to Γ. The boundary condition is a *Neumann condition* while the boundary condition $u=u_0$ on Γ considered previously is said to be a *Dirichlet condition.* In mechanics or physics the Neumann condition (1.37b) corresponds to a given force or flow g on Γ.

We can give the problem (1.37) the following variational formulation (*V*): Find $u\in H^1(\Omega)$ such that

(1.38) $\quad a(u, v)=(f, v)+\langle g, v\rangle \qquad \forall v\in H^1(\Omega)$,

where

$$a(u, v)=\int_\Omega[\nabla u\cdot\nabla v+uv]dx,\ (f, v)=\int_\Omega fv\,dx,\ \langle g, v\rangle=\int_\Gamma gv\,ds.$$

This is equivalent to the following minimization formulation (*M*): Find $u\in H^1(\Omega)$ such that $F(u)\leqslant F(v)$, $\forall v\in H^1(\Omega)$, where

$$F(v)=\frac{1}{2}a(v, v)-(f, v)-\langle g, v\rangle.$$

To see that (1.38) follows from (1.37) we multiply (1.37a) with the test function $v\in H^1(\Omega)$ and integrate over Ω. According to Green's formula (1.18), we then get, since $\frac{\partial u}{\partial n}=g$ on Γ,

$$(f, v)=\int_\Omega(-\Delta u+u)v\,dx=-\int_\Gamma\frac{\partial u}{\partial n}v\,dx+\int_\Omega\nabla u\cdot\nabla v dx+\int_\Omega uv\,dx=$$

$$=-\langle g, v\rangle+\int_\Omega[\nabla u\cdot\nabla v+uv]dx=a(u, v)-\langle g, v\rangle,$$

which proves (1.38).

Let us now also motivate why a solution $u \in H^1(\Omega)$ of the variational problem (1.38) also should satisfy (1.37). Using Green's formula again we find from (1.38) that if u is sufficiently regular, then

$$(f, v)+\langle g, v\rangle=a(u, v)=\int_\Gamma \frac{\partial u}{\partial n} v\, dx+\int_\Omega(-\Delta u+u)v\, dx,$$

so that, rearranging terms,

$$\int_\Omega(-\Delta u+u-f)v\, dx+\int_\Gamma(\frac{\partial u}{\partial n}-g)v\, ds=0 \qquad \forall v\in H^1(\Omega). \tag{1.39}$$

Now, as (1.39) holds in particular for all v in $H_0^1(\Omega)$ and for these functions the boundary term vanishes, we conclude that (1.37a) holds, ie,

$$-\Delta u+u-f=0 \quad \text{in } \Omega.$$

Thus (1.39) is reduced to

$$\int_\Gamma(\frac{\partial u}{\partial n}-g)v\, ds=0 \qquad \forall v\in H^1(\Omega).$$

But varying now v over $H^1(\Omega)$, which means that v will vary freely on Γ, we finally get

$$\frac{\partial u}{\partial n}-g=0 \qquad \text{on } \Gamma,$$

and (1.37b) follows.

We note that the Neumann condition (1.37b) does not appear explicitly in the variational formulation (V); the solution u of (V) is only required to belong to $H^1(\Omega)$ and is not explicitly required to satisfy (1.37b). This boundary condition is instead implicitly contained in (1.38); by first varying v "inside" Ω we obtain (1.37a) and then (1.37b) by varying v on the boundary Γ. Such a boundary condition, that does not have to be explicitly imposed in the variational formulation, is said to be a *natural boundary condition.* This is in contrast to a so-called *essential boundary condition,* like the Dirichlet condition $u=0$ on Γ in eg (1.32), that has to be explicitly satisfied in a variational formulation of the form (1.33).

Let us now formulate a finite element method for the Neumann problem (1.37). Let then T_h be a triangulation of Ω as in Section 1.4 and define

$$V_h=\{v: v \text{ is continuous on } \Omega,\ v|_K \text{ is linear } \forall K\in T_h\}.$$

As parameters to describe the functions in V_h we of course choose the values at the nodes, now including also the nodes on the boundary Γ. Note that the functions in V_h are not required to satisfy any boundary condition and that $V_h \subset H^1(\Omega)$. By starting from (1.38) we now have the following finite element method for (1.37): Find $u_h \in V_h$ such that

$$(1.40) \qquad a(u_h, v) = (f, v) + \langle g, v \rangle \qquad \forall v \in V_h.$$

As in Section 1.4 we see that this problem has a unique solution u_h that can be determined by solving a symmetric, positive definite linear system of equations. We also have the following error estimate

$$\|u - u_h\|_{H^1(\Omega)} \leq \|u - v\|_{H^1(\Omega)} \qquad \forall v \in V_h,$$

and hence as above

$$\|u - u_h\|_{H^1(\Omega)} \leq Ch,$$

if u is regular enough. The function u_h will satisfy the Neumann condition (1.37b) approximatly, i e, $\frac{\partial u_h}{\partial n}$ will be an approximation to g on Γ (cf Problem 1.16).

Remark When formulating a difference method for (1.37) one meets severe difficulties due to the boundary condition (1.37b) unless Ω has a very simple shape such as a rectangle. On the other hand, in the finite element formulation, the same boundary condition does not cause any complication. □

Problems

1.16 Show that the problem

$$-u'' = f \qquad \text{on } I = (0, 1),$$
$$u(0) = u'(1) = 0,$$

can be given the following variational formulation: Find $u \in V$ such that

$$(u', v') = (f, v) \qquad \forall v \in V,$$

where $V = \{v \in H^1(I) : v(0) = 0\}$. Formulate a finite element method for this problem using piecewise linear functions. Determine the corresponding linear system of equations in the case of a uniform partition and study in particular how the boundary condition $u'(1) = 0$ is approximated by the method.

1.17 Show that the problems (*M*) and (*V*) of this section are equivalent.

1.18 Let Ω be a bounded domain in the plane and let the boundary Γ of Ω be divided into two parts Γ_1 and Γ_2. Give a variational formulation of the following problem:

$$\begin{aligned} \Delta u &= f && \text{in } \Omega, \\ u &= u_0 && \text{in } \Gamma_1, \\ \frac{\partial u}{\partial n} &= g && \text{on } \Gamma_2, \end{aligned}$$

where f, u_0 and g are given functions. Then formulate a finite element method for this problem. Also give an interpretation of this problem in mechanics or physics.

1.19 Consider the finite element method (1.2) for the model problem (1.29). Let $G_i \in H_0^1(I)$ satisfy

$$(1.41)\quad (v', G_i') = v(x_i) \qquad \forall v \in H_0^1(I),$$

where x_i is a given node, $i=1, \ldots, M$. Prove that G_i is given by

$$G_i(x) = \begin{cases} (1-x_i)x & \text{for } 0 \leqslant x \leqslant x_i, \\ x_i(1-x) & \text{for } x_i \leqslant x \leqslant 1. \end{cases}$$

Note that G_i is the *Green's function* for (1.29) associated with a delta function $\delta(x_i)$ at node x_i (G_i satisfies $-G_i'' = \delta(x_i)$ on I, $G_i(0) = G_i(1) = 0$). Further, note that it so happens that $G_i \in V_h$. Now, by choosing $v = e = u - u_h$ in (1.41), show that

$$e(x_i) = (e', G_i') = 0, \qquad i = 1, \ldots, M.$$

Thus, u_h is in fact exactly equal to u at the node points x_i. This somewhat surprising fact is a true one-dimensional effect due to the fact that the Green's function $G_i \in V_h$, and does not exist in higher dimensions. The technique of working with a Green's function in this way is however useful in proving for instance pointwise error estimates (maximum norm estimates) in higher dimensions.

1.8 Remarks on programming

Let us briefly discuss some of the essential features of a typical computer program implementing a finite element method. To be concrete we consider the Neumann problem of the previous section. Thus, let $T_h = \{K\}$ be a triangulation of the domain $\Omega \subset R^2$ with boundary Γ and let V_h be the corresponding space of continuous piecewise linear functions. Let N_i, $i=1,\ldots, M$, denote the nodes of T_h and $\varphi_1,\ldots, \varphi_M$ the natural base for V_h, ie, $\varphi_i(N_j)=\delta_{ij}$. We want to find the solution $\xi \in R^M$ of the linear system of equations

$$A\xi=b, \tag{1.42}$$

where $A=(a_{ij})$, $b=(b_1,\ldots, b_M)$,

$$a_{ij}=\sum_{K\in T_h} a_{ij}^K,\quad b_i=\sum_{K\in T_h} b_i^K,$$

$$a_{ij}^K=\int_K \{\nabla\varphi_i\cdot\nabla\varphi_j+\varphi_i\varphi_j]dx, \tag{1.43}$$

$$b_i^K=\int_K f\varphi_i dx+\int_{K\cap\Gamma} g\varphi_i ds.$$

The computer program is naturally divided into subroutines carrying out the following tasks:

(a) Input of data f, g, Ω and coefficients of the equation.
(b) Construction and representation of the triangulation T_h.
(c) Computation of the element stiffness matrices a^K and element loads b^K.
(d) Assembly of the global stiffness matrix A and load vector b.
(e) Solution of the system of equations $A\xi=b$.
(f) Presentation of result.

Let us now consider the steps (b)−(e) in more detail.

(b) Construction and representation of the triangulation T_h

A program for *automatic triangulation* of a given domain may be based on the idea of successive refinement of an initial coarse triangulation; for example, we may refine each triangle by connecting the midpoints of each side (see Fig 1.14).

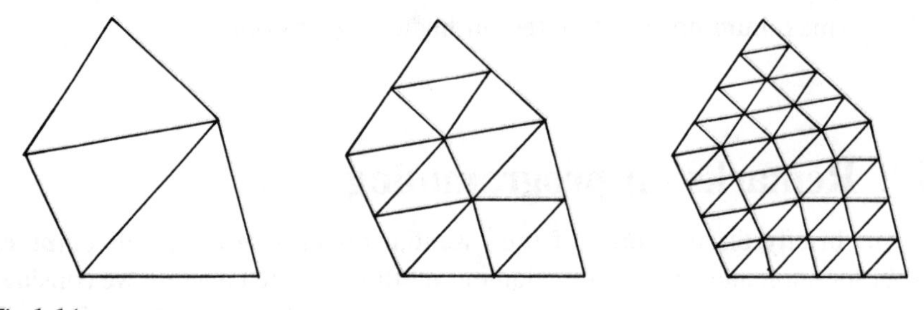

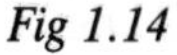

Fig 1.14

A triangulation process of this type leads to *quasi-uniform* meshes where the triangles have essentially the same size in all parts of Ω. If the boundary of Ω is curved, this technique has to be modified close to the boundary.

As discussed below, it is often desirable to be able to construct triangulations where the size of the triangles varies considerably in different parts of Ω. In fact one would need smaller triangles in regions where the exact solution varies quickly or where certain derivatives of the exact solution are large, see Fig 1.16 where the triangles get smaller in the area where the solution has a quick variation (cf Example 1.3). A possible refinement strategy is indicated in Fig 1.15. Here, different coarse grid triangles are refined differently. Notice also the dotted lines introduced to complete the triangulation in the transition zone between regions with elements of different size. Recently, methods which automatically refine triangulations where needed, so-called *adaptive methods,* have been introduced, cf Section 4.6 below.

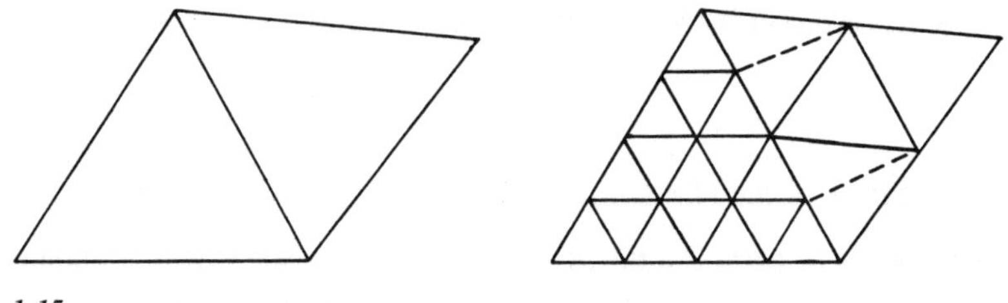

Fig 1.15

To represent a given triangulation T_h one may proceed as follows: Let N_i, $i=1,\ldots, M$, and K_n, $n=1,\ldots, N$ be enumerations of the nodes and triangles of T_h, respectively. Then T_h may be specified using the two arrays $Z(2, M)$ and $T(3, N)$, where $Z(j, i)$, $j=1, 2$, are the coordinates of node N_i and $T(j, n)$, $j=1, 2, 3$, are the number of the vertices of triangle K_n. As an example let us consider the following triangulation where the numbers of the triangles are indicated by a circle:

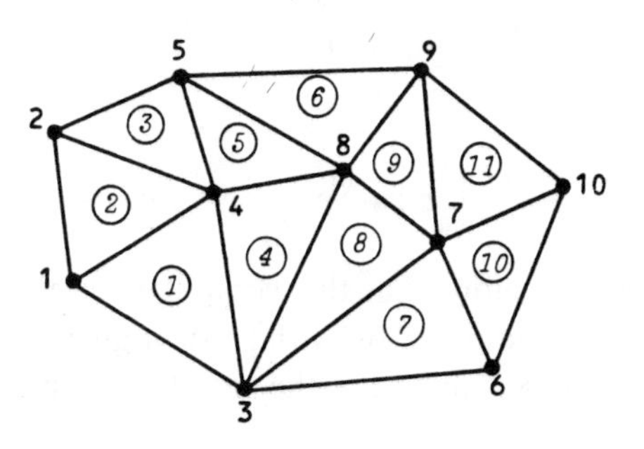

In this case we have

$$T=\begin{bmatrix} 1 & 1 & 2 & 3 & 4 & 5 & 3 & 3 & 8 & 7 & 7 \\ 3 & 4 & 4 & 8 & 8 & 8 & 6 & 7 & 7 & 6 & 10 \\ 4 & 2 & 5 & 4 & 5 & 9 & 7 & 8 & 9 & 10 & 9 \end{bmatrix}.$$

If we want to use Gaussian elimination to solve the system of equations $A\xi=b$ (see Chapter 6), it is important that the nodes are enumerated in a suitable way. For instance, if we intend to store the stiffness matrix A as a band matrix, then we want the band width of A to be (nearly) minimal.

Writing a general program for triangulation, including refinement and node enumeration (if needed), is a complicated task that we will not comment on further. Let us just note that if the geometry of Ω is simple and we are satisfied with a quasi-uniform triangulation, then it is rather easy to write a subroutine for triangulation in each individual case.

We now assume that in some way we have obtained a triangulation T_h and that T_h is represented by the arrays Z and T as above.

(c) Computation of the element stiffness matrices

The next step is to compute the element stiffness matrices with elements a_{ij}^K given by (1.43). We know that $a_{ij}^K \neq 0$ only if both N_i and N_j are nodes of K. Let now $K_n \in T_h$. Then $T(\alpha, n)$, $\alpha=1, 2, 3$, are the numbers of the vertices of K_n, and the x_i-coordinates, $i=1, 2$, for these vertices are given by $Z(i, T(\alpha, n))$, $\alpha=1, 2, 3$. Knowing the vertices of K_n we can now compute the element stiffness matrix $A^{(n)}=(a_{\alpha\beta}^n)$, $\alpha, \beta=1, 2, 3$, for element K_n

$$a_{\alpha\beta}^n=\int_{K_n}[\nabla\psi_\alpha\cdot\nabla\psi_\beta+\psi_\alpha\psi_\beta]dx,$$

where ψ_α is the linear function on K_n that takes the following values:

$$\psi_\alpha(N_{T(\beta, n)})=\begin{cases}1 & \text{if } \alpha=\beta\\ 0 & \text{if } \alpha\neq\beta \quad \alpha, \beta=1, 2, 3.\end{cases}$$

We can also compute

$$b_\alpha^n=\int_{K_n} f\psi_\alpha dx+\int_{\Gamma\cap K_n} g\psi_\alpha, ds, \qquad \alpha=1, 2, 3.$$

Thus, what we need is a subroutine that computes the element stiffness matrix $A^{(n)}=(a_{\alpha\beta}^n)$ and right hand side $b^{(n)}=(b_\alpha^n)$ for a given triangle K_n. We then loop over all elements K_n and store the result on a scratch file.

(d) Assembly of global stiffness matrix

To assemble the global stiffness $A=(a_{ij})$ we just loop over all elements K_n and successively add in the contributions from different K_n as follows (here A(M, M) and b(M) are arrays where the matrix A and right hand side b will be stored):

Set $A(i, j)=0, \quad b(i)=0, \; i, j=1, \ldots, M.$

For $n=1, \ldots, N$, fetch $A^{(n)}=(a^n_{\alpha\beta})$ and $b^{(n)}=(b^n_\alpha)$ from scratch file and set

$$A(T(\alpha, n), T(\beta, n))=A(T(\alpha, n), T(\beta, n))+a^n_{\alpha\beta},$$

$$b(T(\alpha, n))=b(T(\alpha, n))+b^n_\alpha \qquad \alpha, \beta=1, 2, 3.$$

(e) Solution of the linear system $A\xi=b$

To solve $A\xi=b$ we may use various variants of Gaussian elimination or iterative methods. This is discussed in more detail in Chapters 6 and 7.

Remark In practice we do not use an array A(M, M) for the stiffness matrix A; since A is sparse this would not be economical and would require storage of a large number of zero elements. Instead A is stored e g as a band matrix if Gaussian elimination is to be used to solve $A\xi=b$, or if an iterative method is used, then only the nonzero elements of A are stored (see Chapters 6 and 7 below). □

Remark In a certain variant of Gaussian elimination (the frontal method) the assembly and elimination is carried out in parallel which may save storage (cf Section 6.5 below). □

Remark Once the stiffness matrix A for the Neumann problem (1.37) has been determined, for which the functions in V_h do not satisfy any boundary conditions, we may directly derive the systems of equations $\tilde{A}\xi=\tilde{b}$ corresponding to other boundary conditions. If on a part Γ_1 of the boundary Γ we replace the Neumann condition $\frac{\partial u}{\partial n}=g$ with the Dirichlet boundary condition $u=u_0$ on Γ_1, then we obtain the corresponding system $\tilde{A}\xi=\tilde{b}$ by simply deleting the rows in A corresponding to the nodes on Γ_1 and by entering the values of ξ given by the Dirichlet boundary condition. □

1.9 Remarks on finite element software

Writing a finite element program for a general class of problems with general geometry and variable coefficients (cf Example 2.7 below) is very time consuming and requires expert knowledge. Therefore, much effort may be saved by using, at least in part, existing software. There are several general purpose finite element codes available for academic or commercial use. In particular let us mention the codes with which we have some experience, namely CLUB MODULEF based at INRIA in France [CM] which is an extensive general purpose library of finite element routines, FIDAP (Fluid Dynamics Analysis Package) by M.S. Engelman [Fi] for problems in fluid mechanics, the adaptive multigrid code for elliptic and parabolic problems PLTMG (Piecewise Linear Triangular Multi Grid) by R. Bank [Ba], the smaller LSD/FEM package by M. Bercovier [Be] and the MACFEM program for the Macintosh personal computer by O. Pironneau [Pi]. These codes have a modular structure, clear documentation, give access to the source code and thus are suitable for research, development and educational purposes.

Problem

1.20 Write a computer program implementing the ideas of Section 1.8. Assume first simple geometry, e g Ω a square, and a uniform triangulation. Use a standard routine to solve $A\xi=b$ with Gaussian elimination and A stored as a band matrix.

Example 1.3 Consider the Poisson equation (1.16) in a disc with radius 1 centered at the origin and with the load $f=-1$ in a small disc with radius 0.25 centered at (0.5, 0.5), and f equal to zero elsewhere. In Fig 1.16 we give the finite element mesh together with the level curves and the graph of the corresponding finite element solution obtained by applying a modification of the adaptive PLTMG-code [Ba] to this problem, see [EJ2], [E]. PLTMG uses piecewise linears on triangles, and thus corresponds to (1.20), and also automatically refines the finite element mesh in order to control the error in a chosen norm. We notice that the elements are smaller in the area where the solution has a quick variation, cf Section 4.6 below.

For more information on adaptive methods, see Section 4.6. Note that also the triangulation on the cover was generated by the modification of PLTMG applied to the Laplace equation with Dirichlet boundary conditions in a case where the exact solution has a singularity at the origin, and where accordingly the finite element mesh is refined. More precisely, in this case the exact solution is given in polar coordinates (r, θ) by

$$u(r, \theta)=r^{\gamma}\sin(\gamma\theta), \quad \gamma=\frac{2}{7}. \quad \square$$

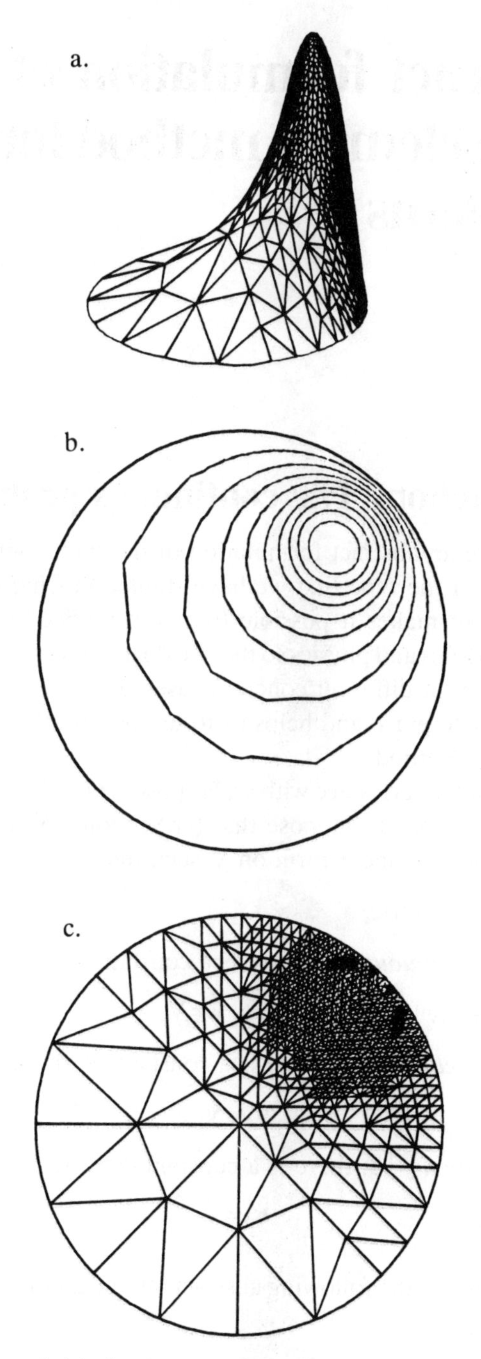

Fig 1.16 Solution graph (a), level curves (b) and triangulation (c) for finite element method for Dirichlet problem.

2. Abstract formulation of the finite element method for elliptic problems

2.1 Introduction. The continuous problem

We shall now give an abstract formulation of the finite element method for elliptic problems of the type that we have studied in Chapter 1. This is not a goal in itself, but makes it possible to give a unified treatment of many problems in mechanics and physics so that we do not have to repeat in principle the same argument in different concrete cases. Further the abstract formulation is very easy to grasp and helps us to understand the basic structure of the finite element method.

Thus, let V be a Hilbert space with scalar product $(.,.)_V$ and corresponding norm $\|\cdot\|_V$ (the V-norm). Suppose that (cf Section 1.5) $a(.,.)$ is a bilinear form on $V\times V$ and L a linear form on V such that

(i) $a(.,.)$ is symmetric,

(ii) $a(.,.)$ is *continuous,* i e, there is a constant $\gamma>0$ such that

(2.1) $$|a(v, w)|\leqslant\gamma\|v\|_V\|w\|_V \qquad \forall v, w\in V,$$

(iii) $a(.,.)$ is V-*elliptic,* i e, there is a constant $\alpha>0$ such that

(2.2) $$a(v, v)\geqslant\alpha\|v\|_V^2 \qquad \forall v\in V.$$

(iv) L is *continuous,* i e, there is a constant $\Lambda>0$ such that

(2.3) $$|L(v)|\leqslant\Lambda\|v\|_V \qquad \forall v\in V.$$

Let us now consider the following abstract minimization problem (M): Find $u\in V$ such that

(2.4) $$F(u)=\min_{v\in V} F(v),$$

where

$$F(v)=\frac{1}{2}\,a(v,v)-L(v),$$

and consider also the following abstract variational problem (*V*): Find $u\in V$ such that

$$\text{(2.5)} \qquad a(u,v)=L(v) \qquad \forall v\in V.$$

Let us now first prove:

Theorem 2.1 The problems (2.4) and (2.5) are equivalent, i e, $u\in V$ satisfies (2.4) if and only if u satisfies (2.5). Moreover, there exists a unique solution $u\in V$ of these problems and the following stability estimate holds

$$\text{(2.6)} \qquad \|u\|_V\leqslant\frac{\Lambda}{\alpha}.$$

Proof Existence of a solution follows from the Lax-Milgram theorem which is variant of the Riesz' representation theorem in Hilbert space theory (see e g [Ne], [Ci], cf also Theorem 13.1 below). The reader unfamiliar with these concepts may simply bypass this remark. To prove that (2.4) and (2.5) are equivalent, we argue exactly as in Section 1.1. We first show that if $u\in V$ satisfies (2.4), then also (2.5) holds, and we leave the proof of the reverse implication to the reader. Thus, let $v\in V$ and $\varepsilon\in R$ be arbitrary. Then $(u+\varepsilon v)\in V$ so that since u is a minimum,

$$F(u)\leqslant F(u+\varepsilon v) \qquad \forall\varepsilon\in R.$$

Using the notation $g(\varepsilon)=F(u+\varepsilon v)$, $\varepsilon\in R$, we thus have

$$g(0)\leqslant g(\varepsilon) \qquad \forall\varepsilon\in R,$$

so that g has a minimum at $\varepsilon=0$. Hence $g'(0)=0$ if the derivative $g'(\varepsilon)$ exists at $\varepsilon=0$. But

$$\begin{aligned} g(\varepsilon)&=\frac{1}{2}\,a(u+\varepsilon v,\ u+\varepsilon v)-L(u+\varepsilon v)\\ &=\frac{1}{2}a(u,\ u)+\frac{\varepsilon}{2}\,a(u,\ v)+\frac{\varepsilon}{2}\,a(v,\ u)+\frac{\varepsilon^2}{2}a(v,\ v)-L(u)-\varepsilon L(v)\\ &=\frac{1}{2}a(u,\ u)-L(u)+\varepsilon a(u,\ v)-\varepsilon L(v)+\frac{\varepsilon^2}{2}a(v,\ v), \end{aligned}$$

where we used the symmetry of $a(.\,,.)$. It follows that

$$0=g'(0)=a(u,\ v)-L(v),$$

which proves (2.5). To prove the stability result we choose $v=u$ in (2.5) and use (2.2) and (2.3) to obtain

$$\alpha\|u\|_V^2 \leq a(u, u) = L(u) \leq \Lambda\|u\|_V,$$

which proves (2.6) upon division by $\|u\|_V \neq 0$. Finally, the uniqueness follows from the stability estimate (2.6) since if u_1 and u_2 are two solutions so that $u_i \in V$ and

$$a(u_i, v) = L(v) \qquad \forall v \in V,\ i=1, 2,$$

then by subtraction we see that $u_1 - u_2 \in V$ satisfies

$$a(u_1 - u_2, v) = 0 \qquad \forall v \in V.$$

Applying the stability estimate to this situation (with $L \equiv 0$, ie, $\Lambda = 0$) we conclude that $\|u_1 - u_2\|_V = 0$, ie, $u_1 = u_2$. □

Remark 2.1 Even without the symmetry condition (i) and with only (ii)–(iv) satisfied, one can prove that there exists a unique $u \in V$ such that

$$a(u, v) = L(v) \qquad \forall v \in V,$$

and the stability estimate (2.6) of course holds (cf Example 2.6 below). In this case there is however no associated minimization problem. □

2.2 Discretization. An error estimate

Now let V_h be a finite-dimensional subspace of V of dimension M. Let $\{\varphi_1, \ldots, \varphi_M\}$ be a basis for V_h, so that $\varphi_i \in V_h$ and any $v \in V_h$ has the unique representation

$$(2.7) \qquad v = \sum_{i=1}^{M} \eta_i \varphi_i, \quad \text{where } \eta_i \in R.$$

We can now formulate the following discrete analogues of the problems (*M*) and (*V*): Find $u_h \in V_h$ such that

$$(2.8) \qquad F(u_h) \leq F(v) \qquad \forall v \in V_h,$$

or equivalently: Find $u_h \in V_h$ such that

$$(2.9) \qquad a(u_h, v) = L(v) \qquad \forall v \in V_h.$$

As in Section 1.2 we see that (2.9) is equivalent to

$$a(u_h, \varphi_j)=L(\varphi_j),\ j=1,\ldots, M.$$

Using the representation

$$(2.10)\qquad u_h=\sum_{i=1}^{M} \xi_i\varphi_i,\quad \xi_i\in R,$$

(2.9) can be written as

$$\sum_{i=1}^{M} a(\varphi_i,\varphi_j)\xi_i=L(\varphi_j),\ j=1,\ldots, M,$$

or, in matrix form,

$$(2.11)\qquad A\xi=b,$$

where $\xi=(\xi_i)\in R^M$, $b=(b_i)\in R^M$ with $b_i=L(\varphi_i)$, and $A=(a_{ij})$ is an $M\times M$ matrix with elements $a_{ij}=a(\varphi_i,\varphi_j)$. From the representation (2.7), we have

$$a(v,v)=a(\sum_{i=1}^{M} \eta_i\varphi_j, \sum_{j=1}^{M} \eta_j\varphi_j)=\sum_{i,\,j=1}^{M} \eta_i a(\varphi_i,\varphi_j)\eta_j=\eta\cdot A\eta,$$

$$L(v)=L(\sum_{j=1}^{M} \eta_j\varphi_j)=\sum_{j=1}^{M} \eta_j L(\varphi_j)=b\cdot\eta,$$

where the dot denotes the usual scalar product in R^M:

$$\xi\cdot\eta=\sum_{i=1}^{M} \xi_i\eta_i.$$

It follows that (2.8) may be formulated as

$$(2.12)\qquad \frac{1}{2}\xi\cdot A\xi-b\cdot\xi=\operatorname*{Min}_{\eta\in R^M}\ [\frac{1}{2}\eta\cdot A\eta-b\cdot\eta].$$

We also have, recalling (2.2),

$$\eta\cdot A\eta=a(v,v)\geqslant\alpha\|v\|_V^2>0,$$

if $v\neq 0$, i e, if $\eta\neq 0$. Since also $a(\varphi_i,\varphi_j)=a(\varphi_j,\varphi_i)$, this proves the following result.

Theorem 2.2 The stiffness matrix A is symmetric and positive definite.

We can now prove the following basic result where the equivalence follows as above.

Theorem 2.3 There exists a unique solution $\xi \in R^M$ to the equivalent problems (2.11) and (2.12), i e, there exists a unique solution $u_h \in V_h$ to the equivalent problems (2.8) and (2.9). Further, the following stability estimate holds:

$$(2.13) \qquad \|u_h\|_V \leq \frac{\Lambda}{\alpha}.$$

Proof Since A is positive definite, A is non-singular, which proves existence and uniqueness. The stability estimate follows by choosing $v=u_h$ in (2.9) which gives, using (2.2) and (2.3),

$$\alpha \|u_h\|_V^2 \leq a(u_h,u_h)=L(u_h)\leq \Lambda \|u_h\|_V,$$

from which (2.13) follows upon division by $\|u_h\|_V \neq 0$.

Remark The stability estimate (2.13) for the finite element solution, which is an analogue of the stability estimate (2.6) for the continuous problem, reflects a very important property of the finite element method. In a certain sense it can be viewed as the theoretical basis for the success of the method. □

Let us now prove the following error estimate:

Theorem 2.4 Let $u \in V$ be the solution of (2.5) and $u_h \in V_h$ that of (2.9) where $V_h \subset V$. Then

$$\|u-u_h\|_V \leq \frac{\gamma}{\alpha} \|u-v\|_V \quad \forall v \in V_h.$$

Proof Since $V_h \subset V$ we have from (2.5) in particular

$$a(u,w)=L(w) \qquad \forall w \in V_h,$$

so that after subtracting (2.9),

$$(2.14) \qquad a(u-u_h,w)=0 \qquad \forall w \in V_h.$$

For an arbitrary $v \in V_h$, define $w=u_h-v$. Then $w \in V_h$, $v=u_h-w$ and by (2.2) and (2.14), we have

$$\alpha\|u-u_h\|_V^2 \leq a(u-u_h,\ u-u_h)=a(u-u_h,\ u-u_h)+a(u-u_h,\ w)$$
$$=a(u-u_h,\ u-u_h+w)=a(u-u_h,\ u-v)\leq \gamma\|u-u_h\|_V\|u-v\|_V,$$

where the last inequality follows from (2.1). Dividing by $\|u-u_h\|_V$ we obtain the desired estimate. □

From the abstract qualitative estimate of Theorem 2.4 we may obtain a quantitative estimate by choosing a suitable function $v \in V_h$ and estimating $\|u-v\|_V$. Usually one then chooses $v=\pi_h u$ where $\pi_h u \in V_h$ is a suitable interpolant of u (e g $\pi_h u$ may be the piecewise linear interpolant $\tilde{u}_h$ of Section 1.3). In Chapter 4 we give estimates for the interpolation error $\|u-\pi_h u\|_V$ in a variety of situations.

2.3 The energy norm

By (2.1) and (2.2) it follows that we may introduce a new norm $\|\cdot\|_a$ on V defined by

$$\|v\|_a^2=a(v, v), \quad v \in V.$$

This norm is *equivalent* to the norm $\|\cdot\|_V$, i e, there are positive constants c and C such that

$$(2.15) \qquad c\|v\|_V \leqslant \|v\|_a \leqslant C\|v\|_V \qquad \forall v \in V.$$

More precisely, we may choose $c=\sqrt{\alpha}$ and $C=\sqrt{\gamma}$. The scalar product $(.,.)_a$ corresponding to $\|\cdot\|_a$ is given by

$$(v,w)_a=a(v,w).$$

The norm $\|\cdot\|_a$ is referred to as the *energy norm*. The error equation (2.14) may now be written

$$(u-u_h, v)_a=0 \qquad \forall v \in V_h,$$

from which follows as in Section 1.3 or by the proof of Theorem 2.4, that

$$(2.16) \qquad \|u-u_h\|_a \leqslant \|u-v\|_a \qquad \forall v \in V_h,$$

or equivalently that u_h is the projection of u onto V_h with respect to the scalar product $(.,.)_a$ (cf Section 1.6). Clearly (2.16) shows that u_h is a best approximation of u in the energy norm.

2.4 Some examples

Let us now consider some concrete examples of the form (2.5). In Chapter 5 further examples from mechanics and physics will be presented. Let Ω be a bounded domain in R^2 or R^3 with boundary Γ. The coordinates in R^2 and R^3 are denoted by $x=(x_1, x_2)$ and $x=(x_1, x_2, x_3)$.

Example 2.1 Let $V=H^1(\Omega)$, $\Omega\subset R^2$,

$$a(v, w)=\int_\Omega[\nabla v\cdot\nabla w+vw]dx,$$

$$L(v)=\int_\Omega fvdx,$$

where $f\in L_2(\Omega)$ in which case (2.5) is a variational formulation of the Neumann problem (1.37) with $g=0$. Let us verify that the conditions (i)–(iv) above are satisfied. Clearly $a(.,.)$ is a symmetric bilinear form on $V\times V$ and L is a linear form. Further,

$$a(v, v)=\|v\|^2_{H^1(\Omega)}$$

and by Cauchy's inequality

$$a(v,w)\leqslant a(v,v)^{1/2}a(w,w)^{1/2}=\|v\|_{H^1(\Omega)}\|w\|_{H^1(\Omega)},$$

which proves (2.1) and (2.2) with $\alpha=\gamma=1$. Finally

$$|L(v)|\leqslant|\int_\Omega fv\,dx|\leqslant\|f\|_{L_2(\Omega)}\|v\|_{L_2(\Omega)},$$

which proves (2.3) with $\Lambda=\|f\|_{L_2(\Omega)}$. □

Example 2.2 Let $V=H^1_0(I)$, $(I=0, 1)$,

$$a(v, w)=\int_I v'w'dx,\quad L(v)=\int_I fv\,dx,$$

where $f\in L_2(I)$ is given, which corresponds to our introductory boundary value problem (1.30). To verify that (i)–(iv) are satisfied, we first note that $a(.,.)$ is obviously symmetric and bilinear and L is linear and since

$$|a(v,w)|\leqslant\|v'\|_{L_2(I)}\|w'\|_{L_2(I)}\leqslant\|v\|_{H^1(I)}\|w\|_{H^1(I)},$$

we have that $a(.,.)$ is continuous. The continuity of L follows as in Example 2.1 and it thus remains to prove the V-ellipticity (2.2), i e, the inequality

$$\int_I(v')^2dx\geqslant\alpha\left(\int_I v^2dx+\int_I(v')^2dx\right)\qquad\forall v\in H^1_0(I),\tag{2.17}$$

for some positive constant α. We shall prove that

$$\int_I v^2dx\leqslant\int_I(v')^2dx\qquad\forall v\in H^1_0(I),\tag{2.18}$$

from which (2.17) follows with $\alpha=\frac{1}{2}$. Since $v(0)=0$ for $v\in H^1_0(I)$, we have

$$v(x)=v(0)+\int_0^x v'(y)dy=\int_0^x v'(y)dy,$$

so that by Cauchy's inequality

$$|v(x)| \leq \int_0^1 |v'| dy \leq (\int_0^1 dy)^{1/2} (\int_0^1 (v')^2 dy)^{1/2} = (\int_0^1 (v')^2 dy)^{1/2}.$$

Squaring this inequality and then integrating over I we obtain (2.18). We note that the inequality (2.18) does not hold for $v(x) \equiv 1$, in which case the left hand side is 1 and the right hand side 0. Thus we need e g a boundary condition of the form $v(0)=0$ for (2.18) to hold in order to control the norm of the function v by the norm of the derivative v', i e, we need a "fixed point" to start from.

If we choose V_h to consist of piecewise linear functions on I as in Section 1.2, we obtain in this case

$$\|u-u_h\|_{H^1(\Omega)} \leq Ch,$$

if u is smooth enough. □

Example 2.3 Let $V = H_0^1(\Omega)$, $\Omega \subset R^2$,

$$a(v,w) = \int_\Omega \nabla v \cdot \nabla w \, dx, \quad L(v) = \int_\Omega fv dx,$$

where $f \in L_2(\Omega)$, in which case (2.5) is a variational formulation of the Dirichlet problem (1.16) for the Poisson equation. We directly see that (i), (ii) and (iv) are satisfied in this case. Thus, only the V-ellipticity, i e, the inequality

$$\text{(2.19)} \qquad \int_\Omega |\nabla v|^2 dx \equiv a(v,v) \geq \alpha \|v\|_{H^1(\Omega)}^2 \equiv \alpha (\int_\Omega (v^2 + |\nabla v|^2) dx)$$

requires comment. To prove (2.19), it is sufficient to prove that there is a constant C such that

$$\text{(2.20)} \qquad \int_\Omega v^2 dx \leq C \int_\Omega |\nabla v|^2 dx \qquad \forall v \in H_0^1(\Omega),$$

since then (2.19) follows with $\alpha = \frac{1}{C+1}$. The proof of (2.20), which is often referred to as Poincare's inequality, is analogous to the proof of (2.18) (cf Problem 2.1 below). With the V_h of Section 1.4 we obtain the error estimate

$$\|u-u_h\|_{H^1(\Omega)} \leq Ch,$$

if u is sufficiently smooth. □

Example 2.4 Consider the following boundary value problem

$$\text{(2.21a)} \qquad \frac{d^4u}{dx^4} = f \qquad \text{for } x \in I = (0, 1),$$

(2.21b) $\quad u(0)=u'(0)=u(1)=u'(1)=0,$

where $f\in L_2(I)$ (cf Problem 1.5). We introduce the space

$$H^2(I)=\{v\in L_2(I):\ v',\ v''\in L_2(I)\},$$

with norm

$$\|v\|_{H^2(I)}=\Big(\int_I[v^2+(v')^2+(v'')^2]dx\Big)^{1/2},$$

and the space

$$H_0^2(I)=\{v\in H^2(I):\ v(0)=v'(0)=v(1)=v'(1)=0\}$$

with the same norm. The problem (2.21) can now be given the variational formulation: Find $u\in V$ such that

$$a(u,v)=L(v) \qquad \forall v\in V,$$

where $V=H_0^2(\Omega)$,

$$a(v,w)=\int_I v''w''dx,\ \ L(v)=\int_I fv\ dx.$$

We see that the conditions (i), (ii) and (iv) are satisfied. By (2.18) we have for $v\in H_0^2(I)$

$$\int_I v^2dx\leq\int_I (v')^2dx\leq\int_I (v'')^2dx,$$

since $v(0)=v'(0)=0$, which proves that

$$\|v\|^2_{H^2(I)}\leq 3\int_I (v'')^2dx\equiv 3\ a(v,v),$$

and (iii) holds with $\alpha=\frac{1}{3}$. □

We now introduce some notation that will be used below. We define

$$D^\alpha v=\frac{\partial^{|\alpha|}v}{\partial x_1^{\alpha_1}\,\partial x_2^{\alpha_2}},$$

where here $\alpha=(\alpha_1,\alpha_2)$, α_i is a non-negative natural number and $|\alpha|=\alpha_1+\alpha_2$. As an example, a partial derivative of order 2 can then be written as $D^\alpha v$ with $\alpha=(2,0)$, $\alpha=(1,1)$ or $\alpha=(0,2)$, which are the α with $|\alpha|=2$. We now define for $k=1,2,\ldots,$

$$H^k(\Omega)=\{v\in L_2(\Omega):\ D^\alpha v\in L_2(\Omega),\ |\alpha|\leq k\},$$

with norm

$$\|v\|_{H^k(\Omega)}=\Big(\sum_{|\alpha|\leq k}\int_\Omega |D^\alpha v|^2 dx\Big)^{1/2}.$$

Thus the space $H^k(\Omega)$ consists of all functions v on Ω that, together with the partial derivatives $D^\alpha v$ of order $|\alpha|$ at most k, belong to $L_2(\Omega)$. The space $H^k(\Omega)$ is a Hilbert space with the indicated norm and corresponding scalar product. The spaces $H^k(\Omega)$ are examples of so called *Sobolev spaces* named after the Russian mathematician S. L. Sobolev 1908–, cf [Ad].

Example 2.5 Let us now consider a fourth-order problem in a two-dimensional domain Ω, namely the *biharmonic problem:*

(2.22a) $\quad \Delta\Delta u=f \quad$ in Ω,

(2.22b) $\quad u=\frac{\partial u}{\partial n}=0 \quad$ on Γ,

where $\frac{\partial}{\partial n}$ denotes differentiation in the outward normal direction to the boundary Γ. This problem gives a formulation of the Stokes equations in fluid mechanics (cf Problem 5.3) and also models the displacement of a thin elastic plate, clamped at its boundary, under a transversal load (cf Problem 5.4). To give a variational formulation of (2.22), we introduce the space

$$H_0^2(\Omega)=\{v\in H^2(\Omega):\ v=\frac{\partial v}{\partial n}=0 \text{ on } \Gamma\}.$$

Now we multiply (2.22a) with $v\in H_0^2(\Omega)$ and integrate over Ω. By Green's formula as $v=\frac{\partial v}{\partial n}=0$ on Γ, we have

$$\int_\Omega fv dx=\int_\Omega \Delta\Delta u\ v\ dx=$$

$$=\int_\Gamma \frac{\partial}{\partial n}(\Delta u)v\ ds-\int_\Omega \nabla(\Delta u)\cdot\nabla v\ dx=$$

$$=-\int_\Omega \nabla(\Delta u)\nabla v dx=-\int_\Gamma \Delta u\frac{\partial v}{\partial n} ds+\int_\Omega \Delta u\Delta v dx=\int_\Omega \Delta u\Delta v\ dx.$$

We are thus led to the following variational formulation of the biharmonic problem (2.22): Find $u\in V$ such that

$$a(u, v)=L(v) \qquad \forall v\in V,$$

where $v=H_0^2(\Omega)$ and

$$a(u, v)=\int_\Omega \triangle u \ \triangle v \ dx, \ L(v)=\int_\Omega fv \ dx.$$

Again we see directly that (i), (ii) and (iv) are satisfied in this case and the V-ellipticity (iii) can easily be proved using the hints of Problem 2.2 below. In Chapter 3 below we shall construct finite element spaces $V_h \subset H_0^2(\Omega)$. □

Example 2.6 Consider the following problem in a domain $\Omega \subset R^2$:

$$\text{(2.23a)} \qquad -\mu\triangle u+\beta_1 \frac{\partial u}{\partial x_1}+\beta_2\frac{\partial u}{\partial x_2}+u=f \qquad \text{in } \Omega,$$

$$\text{(2.23b)} \qquad u=0 \text{ on } \Gamma,$$

where μ and the β_i are constants with $\mu>0$. This is an example of a stationary *convection-diffusion problem;* the Laplace term corresponds to diffusion with diffusion coefficient μ and the first order derivatives correspond to convection in the direction $\beta=(\beta_1, \beta_2)$. Let us here assume that $\mu=1$ and that the size of $|\beta|$ is moderate (for convection-diffusion problems with $|\beta|/\mu$ large, see Chapter 9). By multiplying (2.23a) by a test function $v\in V=H_0^1(\Omega)$, integrating over Ω and using Green's formula for the Laplace-term as usual, we are led to the following variational formulation of (2.23): Find $u\in V$ such that

$$\text{(2.24)} \qquad a(u, v)=L(v) \qquad \forall v\in V,$$

where

$$a(v, w)=\int_\Omega(\nabla v\cdot \nabla w+(\beta_1 \frac{\partial v}{\partial x_1}+\beta_2\frac{\partial v}{\partial x_2}+v)w)dx, \quad L(v)=\int_\Omega fv \ dx.$$

It is clear that $a(.,.)$ is V-elliptic since if $v\in V$, we have by Green's formula:

$$\int_\Omega(\beta_1\frac{\partial v}{\partial x_1}v+\beta_2 \frac{\partial v}{\partial x_2}v)dx=\int_\Gamma v^2(\beta_1 n_1+\beta_2 n_2)ds-$$

$$-\int_\Omega(v \ \beta_1 \frac{\partial v}{\partial x_1}+v \ \beta_2\frac{\partial v}{\partial x_2})dx=-\int_\Omega(\beta_1 \frac{\partial v}{\partial x_1}v+\beta_2 \frac{\partial v}{\partial x_2}v)dx,$$

ie,

$$\int_\Omega(\beta_1 \frac{\partial v}{\partial x_1}+\beta_2\frac{\partial v}{\partial x_2})v \ dx=0,$$

so that

$$a(v, v)=\int_\Omega [|\nabla v|^2+v^2]dx=||v||^2_{H^1(\Omega)}.$$

Existence of a unique weak solution of (2.23) now follows from Remark 2.1. Starting from (2.24) we may formulate the following finite element method for (2.23): Find $u_h \in V_h$ such that

$$(2.25) \qquad a(u_h, v)=L(v) \qquad \forall v \in V_h,$$

where V_h is a finite-dimensional subspace of V. If $\{\varphi_1, \ldots, \varphi_M\}$ is a basis for V_h we have as above that (2.25) is equivalent to the linear system $A\xi=b$ where $A=(a_{ij})$, $a_{ij}=a(\varphi_i, \varphi_j)$, and $b=(b_i)$, $b_i=(f, \varphi_i)$. Note that in this case the matrix A is *not* symmetric.

By the V-ellipticity it follows that solutions of (2.25) are unique and thus A is non-singular so that $A\xi=b$ admits a unique solution, i e, there exists a unique solution u_h of (2.25). By the same argument as in the proof of Theorem 2.4, we also have the error estimate (here $\alpha=1$):

$$\|u-u_h\|_{H^1(\Omega)} \leq \gamma \|u-v\|_{H^1(\Omega)} \quad \forall v \in V_h. \quad \square$$

Example 2.7 Let u be the temperature in a heat conducting body occupying the domain $\Omega \subset R^3$. We have in the stationary case the following relations:

$$(2.26a) \qquad -q_i=k_i(x) \frac{\partial u}{\partial x_i} \qquad \text{in } \Omega, \ i=1, 2, 3, \text{ (Fourier's law)},$$

$$(2.26b) \qquad \text{div } q=f \qquad \text{in } \Omega \text{ (conservation of energy)},$$

where the q_i denotes the heat flow in the x_i-direction, $k_i(x)$ is the heat conductivity at x in the x_i-direction and f(x) is the heat production at x. If $k_i(x)\equiv 1$, $x \in \Omega$, $i=1, 2, 3$, i e, if the heat conductivity is constant and equal in all directions, then eliminating q in (2.26), we obtain Poisson's equation $-\Delta u=f$ in Ω. With the k_i non-constant, (2.26) is an example of a partial differential equation with *variable coefficients*. However, the coefficients k_i are not assumed to depend on the solution u. If this was the case and the heat conductivities k_i depended on the temperature u, then (2.26) would be an example of a *non-linear* partial differential equation, see Chapter 13 below.

Let us now give a variational formulation of (2.26) which in the usual way can be used to formulate a finite element method for (2.26). This shows that the presence of the variable coefficients k_i do not introduce any difficulties. We complement (2.26a, b) with the following boundary conditions:

$$(2.26c) \qquad u=0 \qquad \text{on } \Gamma_1,$$

$$(2.26d) \qquad -q \cdot n=g \qquad \text{on } \Gamma_2,$$

where $\Gamma=\Gamma_1\cup\Gamma_2$ is a partition of the boundary Γ and n denotes the outward unit normal to Γ. The condition (2.26d) corresponds to a situation where the heat flow is given on Γ_2.

We introduce the space

$$V=\{v\in H^1(\Omega): v=0 \text{ on } \Gamma_1\},$$

multiply (2.26b) by $v\in V$ and integrate over Ω. By Green's formula we then get

$$\int_\Omega fv\,dx=\int_\Omega v \operatorname{div} q\,dx=\int_\Gamma vq\cdot n\,ds-\int_\Omega q\cdot\nabla v\,dx=$$

$$=\int_\Omega \Sigma_{i=1}^3 k_i(x)\frac{\partial u}{\partial x_i}\frac{\partial v}{\partial x_i}\,dx-\int_{\Gamma_2} gv\,ds,$$

where the last equality follows from (2.26a), (2.26d) and the fact that $v=0$ on Γ_1. Thus we are led to the following variational formulation of (2.26): Find $u\in V$ such that

$$a(u,v)=L(v) \qquad \forall v\in V, \tag{2.27}$$

where

$$a(v,w)=\int_\Omega \sum_{i=1}^3 k_i(x)\frac{\partial v}{\partial x_i}\frac{\partial w}{\partial x_i}\,dx,$$

$$L(v)=\int_\Omega fv\,dx+\int_{\Gamma_2} gv\,ds.$$

We easily verify that the conditions (i)−(iv) are satisfied under the following hypothesis: There are positive constants c and C such that

$$c\leqslant k_i(x)\leqslant C,\ x\in\Omega,\ i=1, 2, 3,$$

$f\in L_2(\Omega)$, $g\in L_2(\Gamma_2)$, and the area of Γ_1 is positive.

Starting from (2.27) we may now formulate a finite element method for (2.26) by replacing V by a finite element space $V_h\subset V$. This leads to a linear system $A\xi=b$ with stiffness matrix $A=(a_{ij})$ with elements $a_{ij}=a(\varphi_i,\varphi_j)$ where $\{\varphi_1,\ldots,\varphi_M\}$ is a basis for V_h. To find the a_{ij} we have to compute integrals involving the variable coefficients $k_i(x)$. In practice we may for this purpose want to use numerical quadrature, cf Chapter 12. □

Problems

2.1 Let Ω be a square with side 1. Show that

$$\left(\int_\Omega v^2dx\right)^{1/2}\leqslant\left(\int_\Omega|\nabla v|^2dx\right)^{1/2} \qquad \forall v\in H_0^1(\Omega).$$

2.2 Let Ω be a square with boundary Γ. Show that there is a constant C such that

$$\|v\|^2_{H^2(\Omega)} \leq C\int_\Omega (\Delta v)^2 dx \qquad \forall v \in H^2_0(\Omega),$$

by using the boundary conditions $v = \frac{\partial v}{\partial n} = 0$ on Γ and the fact that by Green's formula,

$$\int_\Omega \left(\frac{\partial^2 v}{\partial x_1 \partial x_2}\right)^2 dx = \int_\Omega \frac{\partial^2 v}{\partial x_1^2} \frac{\partial^2 v}{\partial x_2^2} dx \qquad \forall v \in H^2_0(\Omega).$$

Note that if $v=0$ on Γ, then also $\frac{\partial v}{\partial s} = 0$ on Γ, where $\frac{\partial}{\partial s}$ differentiation in a tangental direction to Γ.

2.3 Give a variational formulation of the problem

$$\frac{d^4 u}{dx^4} = f \qquad \text{for } 0<x<1,$$
$$u(0) = u''(0) = u'(1) = u'''(1) = 0,$$

and show that the conditions (i)−(iv) are satisfied. Which boundary conditions are essential and which are natural? What is the interpretation of the boundary conditions if u represents the deflection of an elastic beam?

2.4 Let Ω be a square with boundary Γ. Show that there is a constant C such that

$$\left(\int_\Gamma v^2 ds\right)^{1/2} \leq C\|v\|_{H^1(\Omega)} \quad \forall v \in H^1(\Omega).$$

Using this result show that the linear functional $L: H^1(\Omega) \to R$ defined by

$$L(v) = \int_\Gamma gv\, ds$$

is continuous if $g \in L_2(\Gamma)$, i e, if $\int_\Gamma g^2 ds < \infty$.

2.5 Give a variational formulation of the inhomogeneous Neumann problem

$$-\Delta u + u = f \qquad \text{in } \Omega,$$
$$\frac{\partial u}{\partial n} = g \qquad \text{on } \Gamma,$$

and check if the conditions (i)–(iv) of Section 2.1 are satisfied. Give an example of a problem in mechanics that takes this form.

2.6 Give a variational formulation of the problem

$$-\Delta u = f \qquad \text{in } \Omega,$$

$$\gamma u + \frac{\partial u}{\partial n} = g \qquad \text{on } \Gamma,$$

where γ is a constant. When are conditions (i)–(iv) satisfied? Give an interpretation of the boundary condition (which is sometimes referred to as a Robin (or third type) boundary condition).

2.7 Consider the variational problem (2.27) with variable coefficients. Suppose that Ω is composed of two parts Ω_1 and Ω_2 with common boundary S (see Fig 2.1) and suppose the coefficients $k_i(x)$ are defined by

$$k_i(x) = \begin{cases} \varkappa_1 & \text{for } x \in \Omega_1, \\ \varkappa_2 & \text{for } x \in \Omega_2, \end{cases}$$

where the $\varkappa_i$ are positive constants.

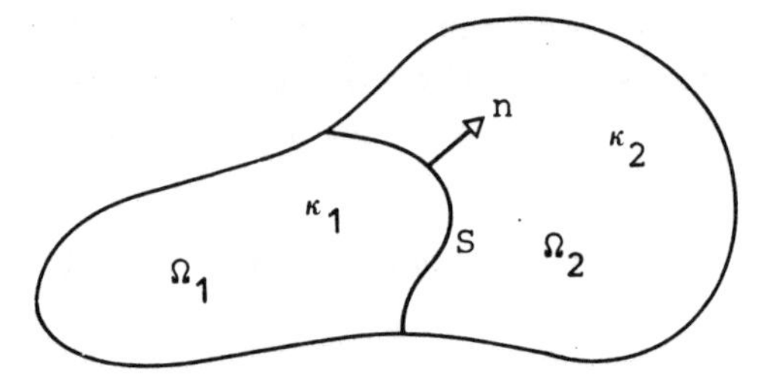

Fig 2.1

In this case (2.27) models stationary heat conduction in an isotropic body composed of two materials with heat conductivity coefficients $\varkappa_1$ and $\varkappa_2$ occupying the regions Ω_1 and Ω_2. Show (formally) that $u \in V$ satisfies (2.27) if and only if

(2.28a) $$-\varkappa_j \Delta u = f \qquad \text{in } \Omega_j,\ j=1, 2,$$

(2.28b) $$u = 0 \qquad \text{on } \Gamma_1,$$

(2.28c) $$q \cdot n = g \qquad \text{on } \Gamma_2,$$

(2.28d) $$\varkappa_1 \frac{\partial u_1}{\partial n} = \varkappa_2 \frac{\partial u_2}{\partial n} \qquad \text{on } S,$$

where $\frac{\partial u_j}{\partial n}$ denotes the derivative of $u_j = u|_{\Omega_j}$ in a direction n normal to S.

Notice that (2.28d) represents a balance of heat flowing between Ω_1 and Ω_2. Observe that this relation is "automatically built in" in the variational formulation (2.27).

2.8 Show (formally) that u is the solution of the variational problem

$$(2.29) \quad \min_{v\in H_0^1(I)} \left[\frac{1}{2}\int_I k(x)(v')^2dx-\int_I v\,dx\right],$$

where $I=(0, 1)$, and

$$k(x)=\begin{cases} 1 & \text{if } x\in I_1=(0, \frac{1}{2}),\\ \frac{1}{2} & \text{if } x\in I_2=(\frac{1}{2}, 1),\end{cases}$$

if and only if u satisfies

$$\begin{aligned} &-k(x)u''(x)=1 \text{ in } I_1 \text{ and } I_2,\\ (2.30) \quad &u_1=u_2,\ 2\frac{du_1}{dx}=\frac{du_2}{dx} \qquad \text{for } x=\frac{1}{2},\\ &u(0)=u(1)=0,\end{aligned}$$

where $u_i=u|_{I_i}$, $i=1, 2$. Then formulate a finite element method for (2.30) using piecewise linear functions. Determine the corresponding linear system in the case of a uniform partition and give an interpretation of this system as a difference method for (2.30).

2.9 Show that if u is the solution of the Dirichlet problem

$$(2.31) \quad \begin{aligned} -\Delta u&=f \qquad \text{in } \Omega,\\ u&=0 \qquad \text{on } \Gamma,\end{aligned}$$

where $f\in L_2(\Omega)$ and $\Omega\subset R^2$, then $p=\nabla u$ is the solution of the minimization problem

$$(2.32) \quad \min_{q\in H_f} \frac{1}{2}\int_\Omega |q|^2dx,$$

where

$$\begin{aligned} H_f&=\{q\in H: \text{div } q+f=0 \text{ in } \Omega\},\\ H&=\{q=(q_1, q_2): q_i\in L_2(\Omega)\}.\end{aligned}$$

The minimization problem (2.32) corresponds to the *Principle of minimum complementary energy* in mechanics. Starting from (2.32), replacing H_f by a finite-dimensional subspace, one may construct finite element methods of so-called *equilibrium* type (for such a method the equilibrium condition div $q+f=0$ will be satisfied exactly in the discrete model). Methods of this type may in certain cases have advantages as compared to the conventional finite element methods, so-called *displacement methods,* that we have studied above (in a displacement method for (2.26) the compatibility relation (2.26a) is satisfied exactly). Hint: First show that $p \in H_f$ is a solution of (2.32) if and only if

$$\int_\Omega p \cdot q \, dx = 0 \qquad \forall q \in H_0,$$

where $H_0 = \{q \in H, \text{ div } q = 0 \text{ in } \Omega\}$.

2.10 Solve Problem 2.3 with the following alternative boundary conditions:

$$u(0) = -u''(0) + \gamma u'(0) = 0, \; u(1) = u''(1) + \gamma u'(1) = 0,$$

where γ is a positive constant. Also give a mechanical interpretation of the boundary conditions.

2.11 Consider the Neumann problem

(2.33a) $-\Delta u = f$ in Ω,

(2.33b) $\dfrac{\partial u}{\partial n} = g$ on Γ,

(2.33c) $\int_\Omega u dx = 0$.

Note that if u satisfies (2.33a, b), then so does $u+c$ for any constant c, and that the condition (2.33c) is added to give uniqueness. Give a variational formulation of (2.33) using the space

$$V = \{v \in H^1(\Omega): \int_\Omega v dx = 0\},$$

and prove that the conditions (i)−(iv) are satisfied.

3. Some finite element spaces

3.1 Introduction. Regularity requirements

We shall now present some commonly used finite element spaces V_h. These spaces will consist of piecewise polynomial functions on subdivisions or "triangulations" $T_h=\{K\}$ of a bounded domain $\Omega\subset R^d$, d=1, 2, 3, into elements K. For d=1, the elements K will be intervals, for d=2, triangles or quadrilaterals and for d=3 tetrahedrons for instance.

We will need to satisfy either $V_h\subset H^1(\Omega)$ or $V_h\subset H^2(\Omega)$, corresponding to second order or fourth order boundary value problems, respectively. Since the space V_h consists of piecewise polynomials, we have

$$V_h\subset H^1(\Omega)\Leftrightarrow V_h\subset C^0(\bar{\Omega}), \tag{3.1}$$

$$V_h\subset H^2(\Omega)\Leftrightarrow V_h\subset C^1(\bar{\Omega}), \tag{3.2}$$

where $\bar{\Omega}=\Omega\cap\Gamma$ and

$$C^0(\bar{\Omega})=\{v: v \text{ is a continuous function defined on } \bar{\Omega}\},$$

$$C^1(\bar{\Omega})=\{v\in C^0(\bar{\Omega}):\ D^\alpha v\in C^0(\bar{\Omega}),\quad |\alpha|=1\}.$$

Thus, $V_h\subset H^1(\Omega)$ if and only if the functions $v\in V_h$ are continuous, and $V_h\subset H^2(\Omega)$ if an only if the functions $v\in V_h$ *and* their first derivatives are continuous. The equivalence (3.1) depends on the fact that the functions v in V_h are polynomials on each element K so that if v is continuous across the common boundary of adjoining elements, then the first derivatives $D^\alpha v$, $|\alpha|=1$, exist and are piecewise continuous so that $v\in H^1(\Omega)$. On the other hand, if v is not continuous across a certain inter-element boundary, i e $v\notin C^0(\bar{\Omega})$, then the derivatives $D^\alpha v$, $|\alpha|=1$, do not exist as functions in $L_2(\Omega)$ and thus $v\notin H^1(\Omega)$ (if v is discontinuous across an element side S, then $D^\alpha v$, $|\alpha|=1$, would be a δ-function supported by S which is not a square-integrable function). In a similar way we realize that (3.2) holds.

To define a finite element space V_h we will have to specify:

(a) the triangulation $T_h=\{K\}$ of the domain Ω,

(b) the nature of the functions v in V_h on each element K (e g linear, quadratic, cubic, etc),

(c) the parameters to be used to describe the functions in V_h.

3.2 Some examples of finite elements

Let us now consider some examples. We first consider the case when Ω is a domain in the plane R^2 with polygonal boundary Γ. Let $T_h=\{K\}$ be a given triangulation of Ω according to Section 1.4 into triangles K. We shall use the following notation for r=0, 1, 2,. . .,

$$P_r(K)=\{v:v \text{ is a polynomial of degree} \leq r \text{ on } K\}.$$

Thus, $P_1(K)$ is the space of linear functions defined on K, i e, functions of the form

$$v(x)=a_{00}+a_{10}x_1+a_{01}x_2,\ x\in K,$$

where the $a_{ij}\in R$. We see that $\{\psi_1, \psi_2, \psi_3\}$, where

$$\psi_1(x)\equiv 1,\ \psi_2(x)=x_1,\ \psi_3(x)=x_2,$$

is a basis for $P_1(K)$, and that dim $P_1(K)=3$, where dim W denotes the dimension of the linear space W.

Further, $P_2(K)$ is the space of *quadratic* functions on K, i e, functions of the form

$$v(x)=a_{00}+a_{10}x_1+a_{01}x_2+a_{20}x_1^2+a_{11}x_1x_2+a_{02}x_2^2,\ x\in K,$$

where the $a_{ij}\in R$. We see that $\{1, x_1, x_2, x_1^2, x_1x_2, x_2^2\}$ is a basis for $P_2(K)$ and that dim $P_2(K)=6$. In general we have

$$P_r(K)=\{v : v(x)=\sum_{0\leq i+j\leq r} a_{ij}x_1^ix_2^j \text{ for } x\in K, \text{ where } a_{ij}\in R\},$$

and

$$\dim P_r(K)=\frac{(r+1)\,(r+2)}{2}.$$

Example 3.1 Let

(3.3) $$V_h=\{v\in C^0(\bar{\Omega}):\ v|_K\in P_1(K),\ \forall K\in T_h\},$$

i e, V_h is the space of continuous piecewise linear functions that we have met in Section 1.7. As parameters, or *global degrees of freedom,* to describe the functions in V_h, we choose

(3.4) the values at the node points of T_h,

(including the node points on Γ). Let us now convince ourselves that this is a legitimate choice and show that a function $v \in V_h$ is uniquely determined by the values (3.4). This is of course intuitively quite obvious but let us anyway carry out the argument in detail here, since it will be a model to be used in more complicated situations below. We then first notice that if $K \in T_h$ is a triangle with vertices a^i, $i=1, 2, 3$, then the degrees of freedom for K corresponding to (3.4), i e, *the element degrees of freedom,* are

(3.5) the values at the vertices a^i, $i=1, 2, 3$.

To show that a function $v \in V_h$ is uniquely determined by the degrees of freedom (3.4) it is sufficient to show:

Theorem 3.1 Let $K \in T_h$ be a triangle with vertices $a^i=(a_1^i, a_2^i)$, $i=1, 2, 3$. A function $v \in P_1(K)$ is uniquely determined by the degrees of freedom (3.5), i e, given the values α_i, $i=1, 2, 3$, there is a uniquely determined function $v \in P_1(K)$ such that

(3.6) $$v(a^i)=\alpha_i \qquad i=1, 2, 3.$$

Proof Since $v(x)=c_1x_1+c_2x_2+c_3$ for some constants $c_i \in R$, (3.6) is equivalent to the linear system of equations

(3.7) $$c_1a_1^i+c_2a_2^i+c_3=\alpha_i, \qquad i=1, 2, 3,$$

in the unknowns c_i. This system has a unique solution for given α_i if and only if the determinant detB of the coefficient matrix

$$B=\begin{bmatrix} a_1^1 & a_2^1 & 1 \\ a_1^2 & a_2^2 & 1 \\ a_1^3 & a_2^3 & 1 \end{bmatrix}$$

is different from zero. However by basic linear algebra

(3.8) $$\det B/2=\text{area of } K,$$

and thus $\det B \neq 0$. Hence B is non-singular, which proves the desired result. Since this argument will be used below, we also give a somewhat different version of this proof. We notice first that

$$\dim P_1(K)=\text{number of degrees of freedom } (=3),$$

i e, (3.7) has the same number of unknowns as equations. In this case it follows, again by basic linear algebra, that $\det B \neq 0$ if and only if solutions of (3.7) are unique, or in other words if the only solution of (3.7) with $\alpha_i=0$, $i=1, 2, 3$, is given by $c_i=0$, $i=1, 2, 3$, or formally:

(3.9) If $v \in P_1(K)$ and $v(a^i)=0$, $i=1, 2, 3$, then $v \equiv 0$.

In fact it is easy to prove (3.9) directly without using (3.8), which shows that we do not have to be able to compute $\det B$ in order to prove that $\det B \neq 0$. As we shall see below, this latter method of proof makes it possible to easily prove analogues of Theorem 3.1 for higher order polynomials in which case a direct computation of the determinant of the corresponding coefficient matrix could be very complicated. □

We can now determine the (nodal) basis functions for $P_1(K)$ associated with the degrees of freedom (3.5), i e, the functions $\lambda_i \in P_1(K)$, $i=1, 2, 3$, such that (see Fig 3.1):

$$\lambda_i(a^j)=\delta_{ij}=\begin{cases}1 & \text{if } i=j\\ 0 & \text{if } i\neq j\end{cases} \qquad i, j=1, 2, 3.$$

A function $v(x) \in P_1(K)$ then has the representation

$$(3.10) \qquad v(x)=\sum_{i=1}^{3} v(a_i)\lambda_i(x) \quad x \in K.$$

To determine the basis functions λ_i, we have to solve the system of equations (3.7) for three special choices of right hand side, namely, (1, 0, 0), (0, 1, 0) and (0, 0, 1).

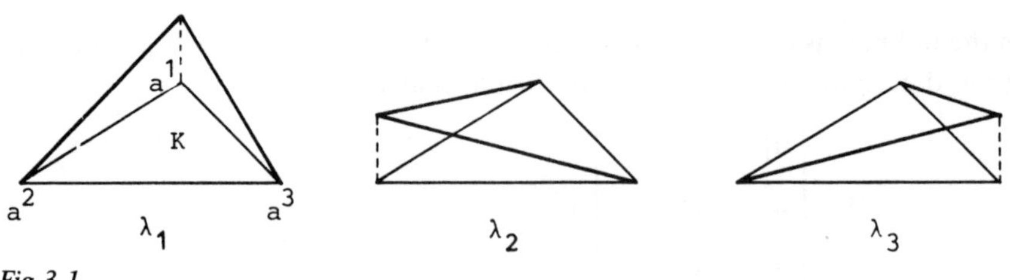

Fig 3.1

The basis function λ_1, say, can also be determined as follows. Let

$$d_1x_1+d_2x_2+d_3=0,$$

be the equation for the straight line through the vertices a^2 and a^3. Then

$$\lambda_1(x)=\gamma(d_1x_1+d_2x_2+d_3),$$

where the constant γ is chosen so that $\lambda_1(a^1)=1$. In the same way we may determine λ_2 and λ_3. If the triangle K has vertices at (1, 0), (0, 1) and (0, 0), then $\lambda_1=x_1$, $\lambda_2=x_2$ and $\lambda_3=1-x_1-x_2$. The notation λ_1, λ_2 and λ_3 for the nodal basis functions for $P_1(K)$ will be kept below.

Given the choice of global degrees of freedom in (3.4), it is natural to describe the space V_h given by (3.3) alternatively as

$$V_h=\{v:\ v|_K\in P_1(K),\ \forall K\in T_h,\ \text{and } v \text{ is continuous at the nodes}\}. \tag{3.11}$$

We then view a function $v\in V_h$ as a piecewise linear function taking on certain values at the nodes of T_h. Let us be careful and check that (3.11) defines the same space as (3.3) above. We need to check if a function $v\in V_h$ according to (3.11) is continuous, i e, if $v\in C^0(\bar{\Omega})$. Clearly, it is sufficient to check that v is continuous across all interelement sides. Thus, let K_1 and K_2 be two triangles in T_h having the common side S with the end points N_1 and N_2, say. Suppose now $v\in V_h$ according to (3.11) and let $v_i=v|_{K_i}\in P_1(K_i)$, i=1, 2, be the restrictions of v to the K_i. Then the function $w=v_1-v_2$ defined on S vanishes at the end points N_1 and N_2 and since w is linear on S it follows that in fact w vanishes on S. Hence, v is continuous across S and we obtain the desired conclusion that $v\in C^0(\bar{\Omega})$.

Example 3.2 Let us now show how to construct a space V_h using piecewise quadratic functions v, i e, $v|_K\in P_2(K)$. Let us first specify the element degrees of freedom. Let $K\in T_h$ be a triangle with vertices a^i, i=1, 2, 3, and denote the midpoints of the sides of K by a^{ij}, i<j, i, j=1, 2, 3, see Fig 3.2.

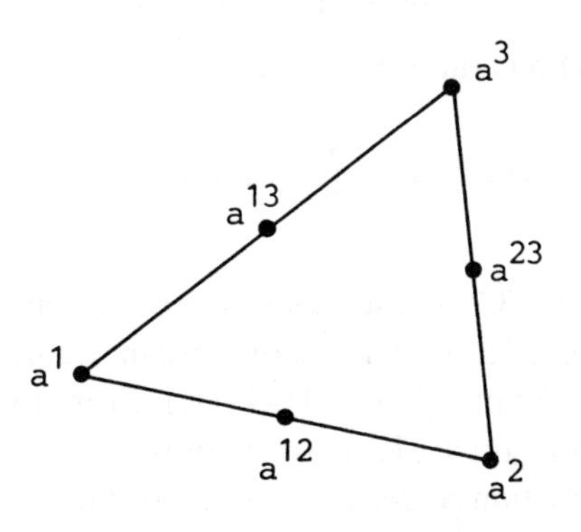

Fig 3.2

We shall prove

Theorem 3.2 A function $v \in P_2(K)$ is uniquely determined by the following degrees of freedom:

$$(3.12) \qquad \begin{array}{ll} v(a^i), & i=1, 2, 3, \\ v(a^{ij}), & i<j,\ i, j=1, 2, 3. \end{array}$$

Proof Since dim $P_2(K)$ is equal to the number of degrees of freedom (=6), it is (see the proof of Theorem 3.1) sufficient to prove that if $v \in P_2(K)$ and

$$(3.13) \qquad v(a^i)=0, \quad v(a^{ij})=0, \quad i<j, \quad i, j=1, 2, 3,$$

then $v \equiv 0$. To this end, consider the side a^2a^3. Along this side the function v has a quadratic variation and v vanishes at the three distinct points a^2, a^{23} and a^3. Thus, (cf Problem 3.1) v vanishes identically on a^{23} which means (cf Problem 3.3) that we can "factor out" the function λ_1 and write

$$v(x)=\lambda_1(x)w_1(x), \qquad x \in K,$$

where $w_1 \in P_1(K)$ and λ_i, $i=1, 2, 3$, are the basis functions for $P_1(K)$ according to Example 3.1. In the same way we see that v also vanishes along the side a^1a^3 which means that we may also factor out the function λ_2, so that

$$v(x)=\lambda_1(x)\lambda_2(x)w_0, \qquad x \in K,$$

where now w_0 has degree zero, i e, $w_0=\gamma=$constant. If we now finally take $x=a^{12}$, we see that

$$0=v(a^{12})=\gamma\lambda_1(a^{12})\lambda_2(a^{12})=\gamma\ \frac{1}{2}\cdot\frac{1}{2},$$

so that $\gamma=0$ and hence $v \equiv 0$ and the proof is complete. □

A function $v \in P_2(K)$ has the representation

$$(3.14) \qquad v= \sum_{i=1}^{3} v(a^i)\lambda_i(2\lambda_i-1)+ \sum_{\substack{i,j=1\\ i<j}}^{3} v(a^{ij})4\lambda_i\lambda_j.$$

To see this, by Theorem 3.2 it is sufficient to check that the right hand side, RH, and left hand side, LH, of (3.14) take the same values at the node points a^i and a^{ij}, since the difference $LH-RH \in P_2(K)$. From (3.14) it is clear what the nodal basis functions for $P_2(K)$ corresponding to the degrees of freedom (3.12) are: the basis function corresponding to a particular degree of freedom, the value at the vertex a^i for instance, is of course the function $\psi \in P_2(K)$ such that $\psi(a^i)=1$ and ψ vanishes at the other five points a^j, a^{ij} (see Fig 3.3).

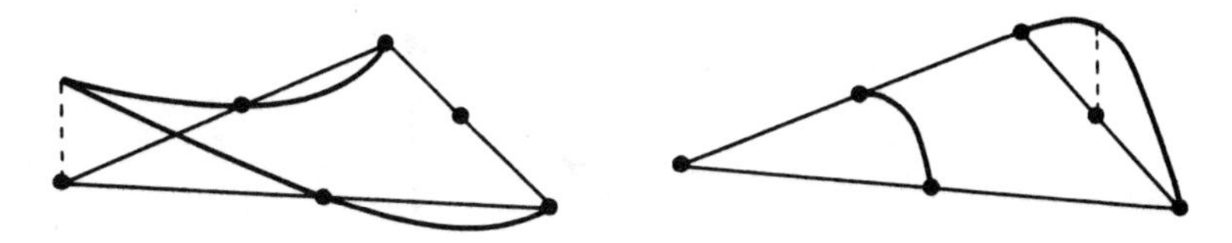

Fig 3.3 Different basis functions for $P_2(K)$

Let us also show that if $v_i \in P_2(K_i)$, $i=1, 2$, where K_1 and K_2 are two triangles with the common side S, and v_1 and v_2 take the same values at the end points and the mid point of S, then v_1 and v_2 agree on S. But this follows immediately from the fact that $w=v_1-v_2$ varies quadratically along S and w vanishes at three distinct points on S so that $w \equiv 0$ on S.

Defining now

$$V_h=\{v \in C^0(\bar{\Omega}): v|_K \in P_2(K), \forall K \in T_h\},$$

we have seen that the global degrees of freedom of the functions $v \in V_h$ can be chosen as follows:

(i) the values of v at the nodes of T_h,

(ii) the values of v at the mid points of all the sides of the triangles in T_h.

The corresponding global basis functions have the following form:

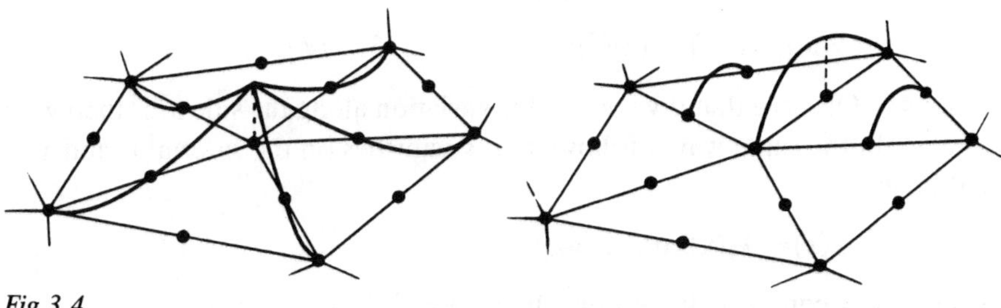

Fig 3.4

Example 3.3 We now define a space V_h using piecewise cubic functions, i e, functions v such that $v|_K \in P_3(K)$, $\forall K \in T_h$. Let K be a triangle with vertices a^i, $i=1, 2, 3$, and define (see Fig 3.5):

$$a^{iij}=\frac{1}{3}(2a^i+a^j), \ i, j = 1, 2, 3, \quad i \neq j,$$

$$a^{123}=\frac{1}{3}(a^1+a^2+a^3).$$

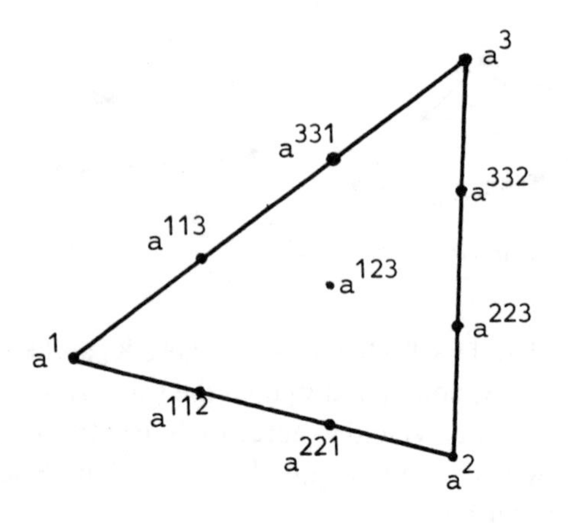

Fig 3.5

We have

Theorem 3.3 A function $v \in P_3(K)$ is uniquely determined by the following degrees of freedom:

$$v(a^i),\ v(a^{iij}),\ i, j=1, 2, 3, \quad i \neq j, \tag{3.15}$$
$$v(a^{123}).$$

Proof Since dim $P_3(K)$ is equal to the number of degrees of freedom (=10), it is sufficient to show that if $v \in P_3(K)$ and

$$v(a^i)=v(a^{iij})=v(a^{123})=0,\ i, j=1, 2, 3, \quad i \neq j, \tag{3.16}$$

then $v \equiv 0$. Observe that if v has a cubic variation along the side a^2a^3 then $v \equiv 0$ on a^2a^3. In the same way it follows that v vanishes on the sides a^1a^3 and a^1a^2 and hence

$$v(x)=\gamma\lambda_1(x)\lambda_2(x)\lambda_3(x),$$

where γ is a constant. If we now choose $x=a^{123}$, we get from (3.16)

$$0=v(a^{123})=\gamma\,\frac{1}{3}\cdot\frac{1}{3}\cdot\frac{1}{3},$$

so that $\gamma=0$ and thus $v \equiv 0$. □

Now let $v_i \in P_3(K_i)$, $i=1, 2$, where K_1 and K_2 are two triangles with common side S and suppose that v_1 and v_2 take the same values at the end points and the two points a^{iij} of S. Since $v_1 - v_2$ varies cubically on S it follows that $v_1 = v_2$ on S (see Fig 3.6).

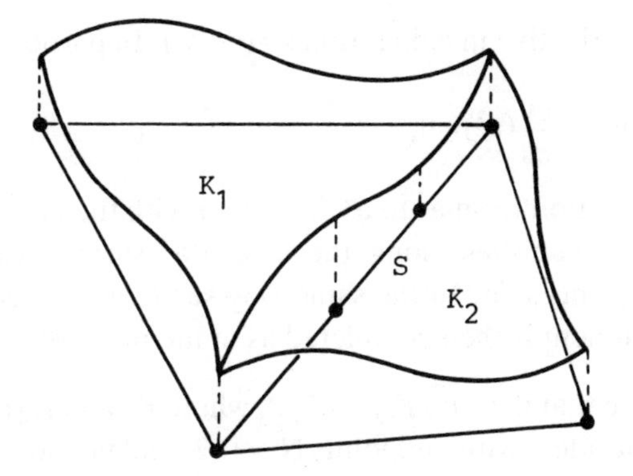

Fig 3.6

We can now introduce the space

$$V_h=\{v\in C^0(\bar{\Omega}): v|_K\in P_3(K),\ \forall K\in T_h\},$$

with the following degrees of freedom:

(i) the values of v at the nodes of T_h.
(ii) the values of v at the points a^{iij} on the sides of T_h,
(iii) the values of v at the center of gravity for all $K\in T_h$.

Example 3.4 There is another way of choosing the degrees of freedom for $P_3(K)$, where K is a triangle with vertices a^i, $i=1, 2, 3$, and center of gravity a^{123}. We have

Theorem 3.4 A function $v\in P_3(K)$ is uniquely determined by the following degrees of freedom:

$$v(a^i), \qquad i=1, 2, 3,$$

$$\frac{\partial v}{\partial x_j}(a^i), \qquad i=1, 2, 3,\ j=1, 2, \tag{3.17}$$

$$v(a^{123}).$$

Proof Since again dim $P_3(K)$ is equal to the number of degrees of freedom, it suffices to prove that if $v\in P_3(K)$ and

$$v(a^i)=\frac{\partial v}{\partial x_j}(a^i)=v(a^{123})=0,\ i=1, 2, 3,\ j=1, 2, \tag{3.18}$$

then $v\equiv 0$. It follows from (3.18) that

$$\frac{\partial v}{\partial s}(a^i)=\frac{\partial v}{\partial x_1}(a^i)s_1+\frac{\partial v}{\partial x_2}(a^i)s_2=0,\ i=1, 2, 3,$$

where $\frac{\partial v}{\partial s}$ is the derivative in a direction $s=(s_1, s_2)$. In particular we then have

$$\frac{\partial v}{\partial s}(a^2)=\frac{\partial v}{\partial s}(a^3)=0,$$

where s is the direction from a^2 to a^3. Together with the fact that $v(a^2)=v(a^3)$ this shows that v vanishes along the side a^2a^3 since v varies as a cubic polynomial along this side. In the same way see that v vanishes on a^1a^2 and a^1a^3 and the argument is then completed as in the proof of Theorem 3.3. □

We further note that if $v_i \in P_3(K_i)$, $i=1, 2$, where K_1 and K_2 are two triangles with the common side S with endpoints $N_j=1, 2$, and v_1 and v_2 agree together with the first derivatives $\frac{\partial v_1}{\partial x_i}(N_j)$ and $\frac{\partial v_2}{\partial x_i}(N_j)$, i, j=1, 2, then $v_1 \equiv v_2$ on S.

The corresponding finite element space $V_h \subset C^0(\bar{\Omega})$ is given by

$$V_h=\{v: v|_K \in P_3(K), \quad \forall K \in T_h, \text{ and } v \text{ and } \frac{\partial v}{\partial x_i}, i=1, 2, \text{ are continuous at the nodes}\},$$

with the following degrees of freedom:

(i) the values of v and $\frac{\partial v}{\partial x_i}$, i=1, 2, at the nodes of T_h,

(ii) the values of v at the center of gravity of each $K \in T_h$. □

Example 3.5 Let us now consider a finite element space V_h satisfying the condition $V_h \subset C^1(\bar{\Omega})$. We will then work with functions that are polynomials of degree five on each triangle; with polynomials of lower degree, special constructions are required to satisfy the C^1-condition.

Theorem 3.5 Let K be a triangle with vertices a^i, i=1, 2, 3 and let a^{ij} be the midpoint on the side a^ia^j, i, j=1, 2, 3, i<j (see Fig 3.2). A function $v \in P_5(K)$ is uniquely determined by the following degrees of freedom:

$$D^\alpha v(a^i),\ i=1, 2, 3,\ |\alpha| \leq 2,$$

$$\frac{\partial v}{\partial n}(a^{ij}),\ i, j=1, 2, 3,\ i<j, \tag{3.19}$$

where $\frac{\partial}{\partial n}$ denotes differentiation in the outward normal direction to the boundary of K.

Proof Since dim $P_5(K)$ is equal to the number of degrees of freedom (=21), it is sufficient as usual to prove that if all the degrees of freedom according to (3.19) are zero, then $v \equiv 0$. To see this, we first note that if s denotes the direction of the side a^2a^3, then

$$v(a^i)=\frac{\partial v}{\partial s}(a^i)=\frac{\partial^2 v}{\partial s^2}(a^i)=0 \qquad i=2, 3. \tag{3.20}$$

Since v is a polynomial on the side a^2a^3 of degree at most 5, it follows that v vanishes on a^2a^3. Further, $\frac{\partial v}{\partial n}$ is a polynomial of degree at most 4 on a^2a^3 and

$$\frac{\partial v}{\partial n}(a^{23})=\frac{\partial v}{\partial n}(a^i)=\frac{\partial}{\partial s}\left(\frac{\partial v}{\partial n}\right)(a^i)=0, \qquad i=2, 3, \tag{3.21}$$

which is only possible if $\frac{\partial v}{\partial n}\equiv 0$ on a^2a^3. Thus, both v and $\frac{\partial v}{\partial n}$ vanish on a^2a^3 which means that we may factor $(\lambda_1(x))^2$ out of v(x) (check this in the special case when a^2a^3 lies on the x_2-axis). Therefore

$$v(x)=(\lambda_1(x))^2 p_3(x), \qquad x \in K,$$

where $p_3 \in P_3(K)$.. In the same way we see that we may also factor out $(\lambda_i(x))^2$, i=2, 3, and thus

$$v=\gamma\lambda_1^2\lambda_2^2\lambda_3^2,$$

where $\gamma \in R$. But $v \in P_5(K)$ and the only possibility then is that $\gamma=0$ so that $v \equiv 0$ on K. □

Now let $v_i \in P_5(K_i)$, i=1, 2, where K_1 and K_2 are two triangles with common side S and suppose that

$$D^\alpha v_1=D^\alpha v_2 \quad \text{at the endpoints of S, } |\alpha|\leq 2,$$

$$\frac{\partial v_1}{\partial n}=\frac{\partial v_2}{\partial n} \quad \text{at the midpoint of S,}$$

where $\frac{\partial}{\partial n}$ denotes differentiation in the normal direction to S. Then we have the relations (3.20) and (3.21) for the difference $w=v_1-v_2$ and it follows that

$$w=\frac{\partial w}{\partial n}=0 \text{ on S.} \tag{3.22}$$

But if w=0 on S we also have that

$$\frac{\partial w}{\partial s}=0 \text{ on S,} \tag{3.23}$$

where $\frac{\partial}{\partial s}$ denotes differentiation in the direction tangential to S. By (3.22) and (3.23) we see the function v defined by $v|_{K_i}=v_i$ varies continuously across S as do its first derivatives.

We may now define the space $V_h \subset C^1(\bar{\Omega})$ as follows

$$V_h=\{v\colon v|_K \in P_5(K),\ \forall K \in T_h,\ D^\alpha v \text{ is continuous at the nodes for } |\alpha| \leq 2 \text{ and } \frac{\partial v}{\partial n} \text{ is continuous at the mid points of each side}\},$$

with the degrees of freedom of (3.19).

Example 3.6 Let us now construct a three-dimensional finite element. We then assume that Ω is the union of a collection $T_h=\{K\}$ of non-overlapping tetrahedrons K such that no vertex of one tetrahedron lies on a side of another tetrahedron. As above, for $r=1, 2, \ldots$, and $K \in T_h$, we define

$$P_r(K)=\{v\colon v \text{ is a polynomial to degree} \leq r \text{ on } K, \text{ i e } v \text{ has the form } v(x)=\sum_{i+j+m\leq r} a_{ijm} x_1^i x_2^j x_3^m,\ a_{ijm} \in R\}.$$

For $r=1$ a function $v \in P_1(K)$ is uniquely determined by the values $v(a^i)$, $i=1, \ldots, 4$, where the a^i are the vertices of K. We can then introduce the space

$$V_h=\{v \in C^0(\bar{\Omega})\colon v|_K \in P_1(K),\ \forall K \in T_h\},$$

and as global degress of freedom we may take the values at the nodes of T_h points. □

Example 3.7 Let us also consider some rectangular finite elements that can be used for example if $\Omega \subset R^2$ is a square. Let then K be a rectangle with vertices a^i, $i=1, \ldots, 4$, and with sides parallel to the coordinate axis in R^2. Define

$$Q_1(K)=\{v\colon v \text{ is } \textit{bilinear} \text{ on } K, \text{ i e, } v(x)=a_{00}+a_{10}x_1+a_{01}x_2+a_{11}x_1x_2,\ x \in K, \text{ where the } a_{ij} \in R\}.$$

It is easy to see (prove this!) that a function $v \in Q_1(K)$ is uniquely determined by the values $v(a^i)$, $i=1, \ldots, 4$. Further, if K_1 and K_2 are two rectangles with the common side S and the functions $v_i \in Q_1(K_i)$ agree at the endpoints of S then $v_1-v_2 \equiv 0$ on S since v_1-v_2 varies linearly on S. We may now define

$$V_h=\{v \in C^0(\bar{\Omega})\colon v|_K \in Q_1(K),\ \forall K \in T_h\}$$

assuming that $T_h=\{K\}$ is a subdivision of Ω into non-overlapping rectangles such that no vertex of any rectangle lies on a side of another rectangle. The values at the nodes may be used as global degrees of freedom.

We can also use polynomials of higher degree on each rectangle. For example we may choose

$$V_h=\{v\in C^0(\bar{\Omega}): v|_K\in Q_2(K),\ \forall K\in T_h\},$$

where $Q_2(K)$ is the set of *biquadratic* functions on K, i e,

$$Q_2(K)=\{v: v(x)=\sum_{i,\, j=0}^{2} a_{ij}x_1^i x_2^j,\ x\in K, \text{ where the } a_{ij}\in R\},$$

and use as global degrees of freedom

(i) the values at the nodes of T_h,
(ii) the values at the midpoints of the sides of T_h,
(iii) the values at the midpoint of each rectangle $K\in T_h$.

Since the use of rectangular elements requires very special geometry of Ω it is of interest to also consider more general quadrilateral elements. The simplest such element is presented in Problem 12.3 below in connection with so-called isoparametric finite elements.

3.3 Summary

We have not yet given a formal definition of what we mean by a "finite element". To fill this gap define a *finite element* to mean a triple (K, P_K, Σ), where

K is a geometric object, for example a triangle,
P_K is a finite-dimensional linear space of functions defined on K,
Σ is a set of degrees of freedom,

such that a function $v\in P_K$ is uniquely determined by the degrees of freedom Σ. From Example 3.1 we have that (K, P_K, Σ), where

K is a triangle,
$P_K=P_1(K)$,
Σ is the values at the vertices of K,

is a finite element. In Fig 3.7 below we have collected some of the most common finite elements (cf [Ci]). The various degrees of freedom are denoted as follows:

. function values

o values of the first derivatives,

O values of the second derivatives,

/ value of the normal derivative,

↗ value of the mixed derivative $\dfrac{\partial^2 v}{\partial x_1 \partial x_2}$.

Finally, Fig 3.8 indicates in the case of two dimensions the *support* of certain basis function $v \in V_h$, i e, the points x such that $v(x) \neq 0$. The different cases correspond to a value at a node, the midpoint of a side or a point in the interior of an element. Clearly the support is always small and if φ and ψ are two basis functions associated with the nodes N_1 and N_2, then the supports of the functions φ and ψ overlap only if N_1 and N_2 belong to the same element.

Degrees of freedom Σ Geometry	Function space P_K	Degree of continuity of corresponding FEM-space V_h
3	$P_1(K)$	C^0
6	$P_2(K)$	C^0
10	$P_3(K)$	C^0
10	$P_3(K)$	C^0
4	$Q_1(K)$	C^0
9	$Q_2(K)$	C^0

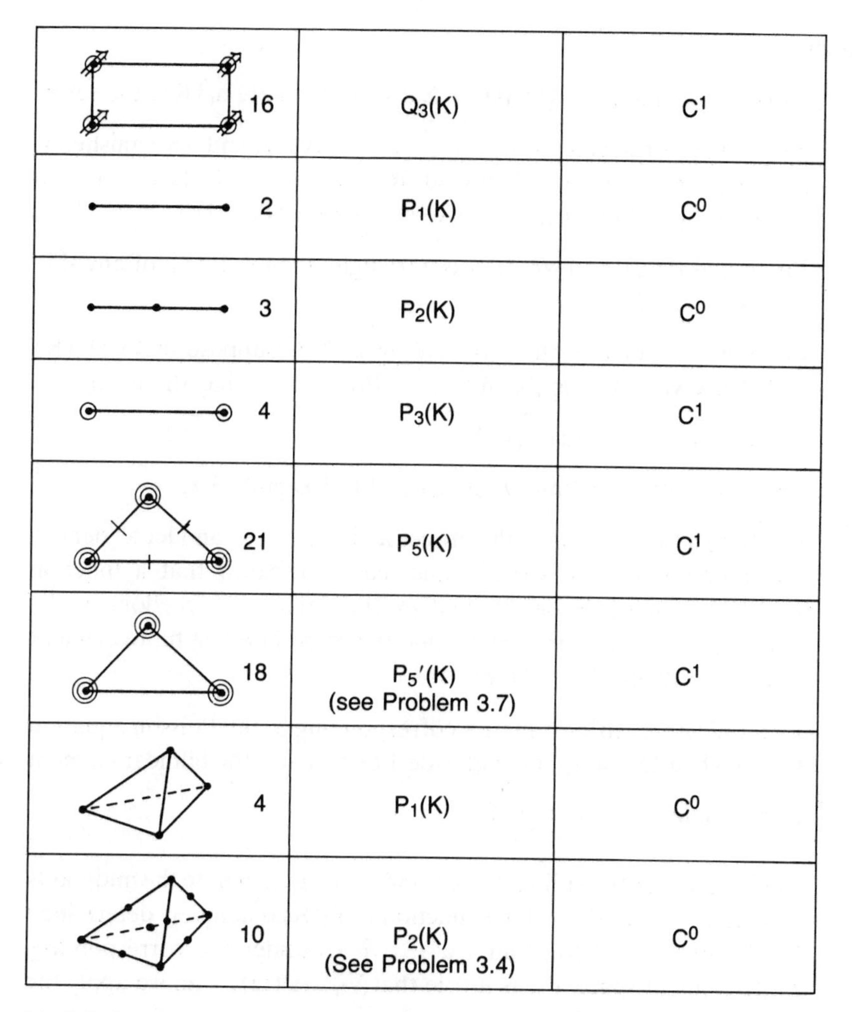

16	$Q_3(K)$	C^1
2	$P_1(K)$	C^0
3	$P_2(K)$	C^0
4	$P_3(K)$	C^1
21	$P_5(K)$	C^1
18	$P_5'(K)$ (see Problem 3.7)	C^1
4	$P_1(K)$	C^0
10	$P_2(K)$ (See Problem 3.4)	C^0

Fig 3.7 Some common finite elements.

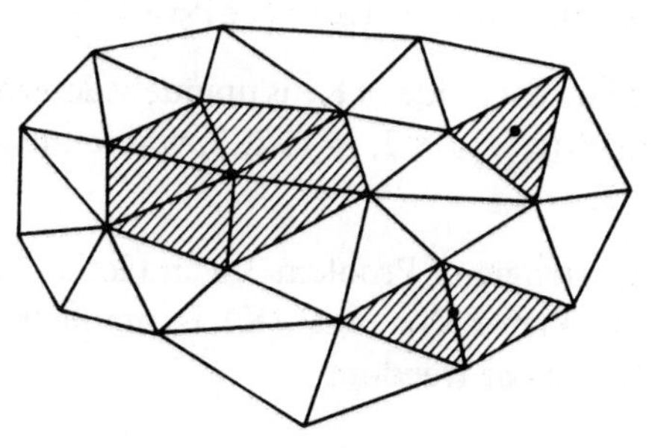

Fig 3.8 The support of different basis functions.

Problems

3.1 Show that if $v \in P_r(I) = \{v: v(x) = \sum_{i=0}^{r} a_i x^i,\ x \in I, \text{ where } a_i \in R\}$, the set of polynomials of degree at most r on the interval I, and if v vanishes at r+1 distinct points on I, then $v \equiv 0$. Recall that if $v \in P_r(I)$ and $v(b)=0$ for some $b \in I$, then $v(x)=(x-b)w(x)$ where $w \in P_{r-1}(I)$.

3.2 Prove that if $v \in P_r(K)$ where K is a triangle, then $v \in P_r(S)$ for any side S of K.

3.3 Let K be a triangle with vertices a^i, i=1, 2, 3. Suppose that $v \in P_r(K)$ and that v vanishes on the side a^2a^3. Prove that v has the form

$$v(x) = \lambda_1(x) w_{r-1}(x), \quad x \in K,$$

where $w_{r-1} \in P_{r-1}(K)$ and λ_1 is defined in Example 3.1.

3.4 Let K be a tetrahedron with vertices a^i, i=1,. . ., 4, and let a^{ij} denote the midpoint on the straight line a^ia^j, $i<j$. Show that a function $v \in P_2(K)$ is uniquely determined by the degrees of freedom: $v(a^i)$, $v(a^{ij})$, i, j=1,. . ., 4, $i<j$. Show that the corresponding finite element space V_h satisfies $V_h \subset C^0(\Omega)$.

3.5 Determine the stiffness matrix corresponding to the Poisson equation (1.16) when Ω is a square with side 1 and we use the bilinear element of Example 3.7 with $h=\frac{1}{4}$.

3.6 Let K be a triangle with vertices a^i and let a^{ij}, $i<j$, denote the midpoints of the sides of K. Show that a function $v \in P_1(K)$ is uniquely determined by the degrees of freedom $v(a^{ij})$, $i<j$. Consider the corresponding finite element space V_h. Is it true that $V_h \subset H^1(\Omega)$? Can we apply the theory of Chapter 2 in this case?

3.7 Show that a function $v \in P_5'(K) = \{v \in P_5(K): \frac{\partial v}{\partial n}$ is a polynomial of degree at most 3 on each side of K$\}$ is uniquely determined by the degrees of freedom $D^\alpha v(a^i)$, $|\alpha| \leq 2$, i=1, 2, 3, where the a^i are the vertices of the triangle K.

3.8 Let K be the triangle of Problem 3.6 and let a^{123} denote the center of gravity of K. Prove that $v \in P_4(K)$ is uniquely determined by the following degrees of freedom

$$v(a^i),\ \frac{\partial v}{\partial x_j}(a^i), \quad i=1, 2, 3,\ j=1, 2,$$

$$v(a^{ij}),\ i, j=1, 2, 3,\ i<j,\ v(a^{123}).$$

Also show that the functions in the corresponding finite element V_h are continuous.

4. Approximation theory for FEM. Error estimates for elliptic problems

4.1 Introduction

For a typical elliptic problem satisfying the conditions (i)–(iv) of Section 2.1, we have by Theorem 2.4

$$\|u-u_h\|_V \leq \frac{\gamma}{\alpha}\|u-v\|_V \qquad \forall v\in V_h.$$

Choosing $v=\pi_h u\in V_h$ to be a suitable interpolant of u and estimating the interpolation error $\|u-\pi_h u\|_V$ we obtain an estimate of the error $\|u-u_h\|_V$. In this chapter we study the problem of estimating the interpolation error $\|u-\pi_h u\|_V$. The interpolant $\pi_h u\in V_h$ is usually chosen so that the degrees of freedom for V_h agree for u and $\pi_h u$. In this case the problem of estimating $\|u-\pi_h u\|_V$ is reduced to the problem of estimating $u-\pi_h u$ individually on each element $K\in T_h$.

4.2 Interpolation with piecewise linear functions in two dimensions

We shall first consider the case where $V=H^1(\Omega)$ and $V_h=\{v\in V: v|_K\in P_1(K), \forall K\in T_h\}$ where $T_h=\{K\}$ is a triangulation of $\Omega\subset R^2$, i e, V_h is the standard finite element space of piecewise linear functions on triangles K (cf Section 1.7). For $K\in T_h$ we define (see Fig 4.1)

$$h_K=\text{the diameter of } K=\text{the longest side of } K,$$
$$\varrho_K=\text{the diameter of the circle inscribed in } K,$$
$$h=\max_{K\in T_h} h_K.$$

To be more precise, we will subsequently be concerned with not only one triangulation T_h but a family of triangulations $\{T_h\}$ that are indexed by the parameter h. We shall below assume that there is a positive constant β independent of the triangulation $T_h \in \{T_h\}$, i e, independent of h, such that

$$\frac{\varrho_K}{h_K} \geqslant \beta \qquad \forall K \in T_h. \tag{4.1}$$

This condition means that the triangles $K \in T_h$ are not allowed to be arbitrarily thin, or equivalently, the angles of the triangles K are not allowed to be arbitrarily small; the constant β is a measure of the smallest angle in any $K \in T_h$ for any $T_h \in \{T_h\}$.

Let N_i, $i=1, \ldots, M$, be the nodes of T_h. Given $u \in C^0(\bar{\Omega})$ we define the interpolant $\pi_h u \in V_h$ by

$$\pi_h u(N_i) = u(N_i) \qquad i=1, \ldots, M.$$

Thus $\pi_h u$ is the piecewise linear function agreeing with u at the nodes of T_h. We will start by estimating the interpolation error $u - \pi_h u$ on each triangle K. We have the following result.

Theorem 4.1 Let $K \in T_h$ be a triangle with vertices a^i, $i=1, 2, 3$. Given $v \in C^0(K)$ let the interpolant $\pi v \in P_1(K)$ be defined by

$$\pi v(a^i) = v(a^i), \; i=1, 2, 3. \tag{4.2}$$

Then

$$\|v - \pi v\|_{L_\infty(K)} \leqslant 2h_K^2 \max_{|\alpha|=2} \|D^\alpha v\|_{L_\infty(K)}, \tag{4.3}$$

$$\max_{|\alpha|=1} \|D^\alpha(v - \pi v)\|_{L_\infty(K)} \leqslant 6 \frac{h_K^2}{\varrho_K} \max_{|\alpha|=2} \|D^\alpha v\|_{L_\infty(K)}, \tag{4.4}$$

where

$$\|v\|_{L_\infty(K)} = \max_{x \in K} |v(x)|.$$

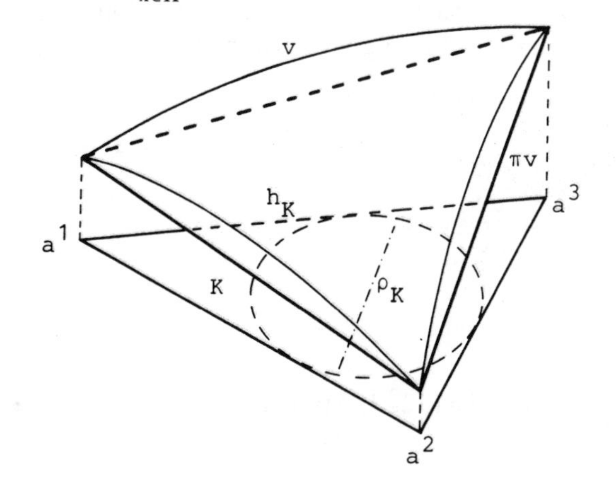

Fig 4.1

Before giving a proof of Theorem 4.1 let us comment on the estimates (4.3) and (4.4). We note that the size of the errors $v-\pi v$ and $D^\alpha(v-\pi v)$ depend on the second partial derivatives of v; the larger these derivatives are, the more "curved" is the surface representing the function v and thus the larger is the deviation $v-\pi v$ from the plane representing πv (see Fig 4.1). Also note that the assumption (4.1) will be used in the estimate (4.4) to bound the quantity h_K/ϱ_K.

Proof of Theorem 4.1 Let λ_i, i=1, 2, 3, be the basis functions for $P_1(K)$ described in Example 3.1. A general function $w \in P_1(K)$ then has the representation

$$w(x)=\sum_{i=1}^{3} w(a^i)\lambda_i(x), \qquad x\in K,$$

so that in particular

$$\pi v(x)=\sum_{i=1}^{3} v(a^i)\lambda_i(x), \qquad x\in K, \tag{4.5}$$

since by (4.2) $\pi v(a^i)=v(a^i)$. We now derive representation formulas for the errors $v-\pi v$ and $D^\alpha(v-\pi v)$, $|\alpha|=1$, using the following Taylor expansion at $x\in K$:

$$v(y)=v(x)+\sum_{j=1}^{2}\frac{\partial v}{\partial x_j}(x)\,(y_j-x_j)+R(x, y),$$

where

$$R(x, y)=\frac{1}{2}\sum_{i, j=1}^{2}\frac{\partial^2 v}{\partial x_i\partial x_j}(\xi)\,(y_i-x_i)\,(y_j-x_j),$$

is the remainder term of order 2 and ξ is a point on the line segment between x and y. In particular by choosing $y=a^i$, we have

$$v(a^i)=v(x)+p_i(x)+R_i(x), \tag{4.6}$$

where

$$p_i(x)=\sum_{j=1}^{2}\frac{\partial v}{\partial x_j}(x)\,(a_j^i-x_j), \quad a^i=(a_1^i, a_2^i),$$

$$R_i(x)=R(x, a^i).$$

Since

$$|a_j^i-x_j|\le h_K, \quad i=1, 2, 3, \quad j=1, 2,$$

we have the following estimate of the remainder term $R_i(x)$:

$$R_i(x) \leq 2h_K^2 \max_{|\alpha|=2} \|D^\alpha v\|_{L_\infty(K)}, \quad i=1,2,3. \tag{4.7}$$

Now (4.5) and (4.6) combine to give

$$\pi v(x) = v(x) \sum_{i=1}^{3} \lambda_i(x) + \sum_{i=1}^{3} p_i(x)\lambda_i(x) + \sum_{i=1}^{3} R_i(x)\lambda_i(x) \qquad x \in K. \tag{4.8}$$

We now need the following lemma whose simple proof is given below.

Lemma 4.1 For j=1, 2 and $x \in K$ we have

$$\sum_{i=1}^{3} \lambda_i(x) = 1, \tag{4.9}$$

$$\sum_{i=1}^{3} p_i(x)\lambda_i(x) = 0, \tag{4.10}$$

$$\sum_{i=1}^{3} \frac{\partial}{\partial x_j} \lambda_i(x) = \frac{\partial}{\partial x_j} \sum_{i=1}^{3} \lambda_i(x) = 0, \tag{4.11}$$

$$\sum_{i=1}^{3} p_i(x) \frac{\partial \lambda_i}{\partial x_j}(x) = \frac{\partial v}{\partial x_j}(x). \tag{4.12}$$

By (4.9), (4.10) and (4.8) we have

$$\pi v(x) = v(x) + \sum_{i=1}^{3} R_i(x)\lambda_i(x),$$

which gives us the following representation of the interpolation error:

$$v(x) - \pi v(x) = -\sum_{i=1}^{3} R_i(x)\lambda_i(x).$$

Since $0 \leq \lambda_i(x) \leq 1$, if $x \in K$, i=1, 2, 3, we can use the previous estimate (4.7) of the remainder term R_i to get

$$|v(x) - \pi v(x)| \leq \sum_{i=1}^{3} |R_i(x)| \; |\lambda_i(x)|$$

$$\leq \max_i |R_i(x)| \sum_{i=1}^{3} \lambda_i(x) \leq 2h_K^2 \max_{|\alpha|=2} \|D^\alpha v\|_{L_\infty(K)}, \qquad x \in K,$$

which proves (4.3).

To prove (4.4) we differentiate (4.5) with respect to x_1 to get

$$\frac{\partial \pi v}{\partial x_1}(x)=\sum_{i=1}^{3} v(a^i)\frac{\partial \lambda_i}{\partial x_1}(x),$$

which together with (4.6) shows that

$$\frac{\partial \pi v}{\partial x_1}(x)=v(x)\sum_{i=1}^{3}\frac{\partial \lambda_i}{\partial x_1}(x)+\sum_{i=1}^{3} p_i(x)\frac{\partial \lambda_i}{\partial x_1}(x)+\sum_{i=1}^{3} R_i(x)\frac{\partial \lambda_i}{\partial x_1}(x). \tag{4.13}$$

Hence, by (4.11) and (4.12) we have

$$\frac{\partial \pi v(x)}{\partial x_1}(x)=\frac{\partial v}{\partial x_1}(x)+\sum_{i=1}^{3} R_i(x)\frac{\partial \lambda_i}{\partial x_1}(x),$$

which gives the following representation of the error $\frac{\partial v}{\partial x_1}-\frac{\partial \pi v}{\partial x_1}$:

$$\frac{\partial v}{\partial x_1}(x)-\frac{\partial \pi v}{\partial x_1}(x)=-\sum_{i=1}^{3} R_i(x)\frac{\partial \lambda_i}{\partial x_1}(x), \qquad x\in K.$$

It is now easy to see (cf Problem 4.2) that

$$\max_{x\in K}\left|\frac{\partial \lambda_i}{\partial x_1}(x)\right|\leqslant\frac{1}{\varrho_K}, \tag{4.14}$$

which together with (4.7) finally gives

$$\left|\frac{\partial v}{\partial x_1}(x)-\frac{\partial \pi v}{\partial x_1}(x)\right|\leqslant 6\frac{h_K^2}{\varrho_K}\max_{|\alpha|=2}\|D^\alpha v\|_{L_\infty(K)}.$$

In the same way we estimate $\frac{\partial v}{\partial x_2}-\frac{\partial \pi v}{\partial x_2}$ and thus (4.4) follows. The proof of the theorem is now complete once the lemma is established.

Proof of Lemma 4.1 The proof is based on the following observation:

$$\pi v=v \text{ if } v\in P_1(K), \tag{4.15}$$

which of course follows from the fact there is a unique function $v\in P_1(K)$ assuming given values at the vertices of K. If we now choose $v(x)\equiv 1$ in (4.8), in which case clearly $v=\pi v$, we get

$$1=\sum_{i=1}^{3}\lambda_i(x), \quad x\in K,$$

since in this case $p_1\equiv R_1\equiv 0$. This proves (4.9) and (4.11) follows directly.

To prove (4.10) we choose $v(x)=d_1x_1+d_2x_2$ in (4.8) with $d_i\in R$. Again $v=\pi v$ and further

$$p_i(x)=d_1(a_1^i-x_1)+d_2(a_2^i-x_2),$$

and $R_i\equiv 0$ so that by (4.8)

$$v(x)=v(x)+\sum_{i=1}^{3}[d_1(a_1^i-x_1)+d_2(a_2^i-x_2)]\lambda_i(x),\quad x\in K.$$

and so for all $d_i\in R$ we have

$$\sum_{i=1}^{3}[d_1(a_1^i-x_1)+d_2(a_2^i-x_2)]\,\lambda_i(x)=0\quad x\in K.$$

This proves (4.10) by choosing $d_i=\frac{\partial v}{\partial x_i}(x)$, $i=1, 2$. Finally, (4.12) follows in a similar way by choosing $v=d_1x_1+d_2x_2$ in (4.13). This finishes the proof of the lemma and the proof of Theorem 4.1 is complete. □

Since Theorem 4.1 states estimates of the interpolation error using the $L_\infty(K)$-norm, it is not ideally suited to give estimates for $\|u-\pi_h u\|_{H^1(\Omega)}$ involving the L_2-norm. For this purpose we will use instead the following analogue of Theorem 4.1. Here we use the following notation for $r=0, 1, 2,\ldots,$

$$|v|_{H^r(\Omega)}=\Big(\sum_{|\alpha|=r}\int_\Omega|D^\alpha v|^2dx\Big)^{1/2}.$$

Note that $|v|_{H^r(\Omega)}$ measures the $L_2(\Omega)$-norm of the partial derivatives of v of order exactly equal to r, whereas derivatives of order less than r are not included. We say that $|\cdot|_{H^r(\Omega)}$ is a *seminorm*. Since we may have $|v|_{H^r(\Omega)}=0$ even if $v\neq 0$ (e g if $v\equiv 1$ and $r\geq 1$), it is not a norm.

Theorem 4.2 Under the assumptions of Theorem 4.1 there is an absolute constant C such that

$$\|v-\pi v\|_{L_2(K)}\leq Ch_K^2|v|_{H^2(K)},$$

$$|v-\pi v|_{H^1(K)}\leq C\,\frac{h_K^2}{\varrho_K}|v|_{H^2(K)}.$$

We see that Theorem 4.1 and 4.2 have exactly the same structure, the only difference being the norm involved, either the L_∞ or the L_2-norm. For simplicity we have chosen to present a proof in the L_∞-case since we then avoid some technical complications (for a proof of Theorem 4.2, see [DS]).

Let us now apply Theorem 4.2 to estimate the global interpolation errors $\|u-\pi_h u\|_{L_2(\Omega)}$ and $|u-\pi_h u|_{H^1(\Omega)}$. We have by summing over $K\in T_h$,

$$\|u-\pi_h u\|^2_{L_2(\Omega)} = \sum_{K \in T_h} \|u-\pi_h u\|^2_{L_2(K)} \leqslant \sum_{K \in T_h} C^2 h_K^4 |u|^2_{H^2(K)}$$

$$\leqslant C^2 h^4 \sum_{K \in T_h} |u|^2_{H^2(K)} = C^2 h^4 |u|^2_{H^2(\Omega)},$$

and similarly using (4.1), i e, $\dfrac{h_K}{\varrho_K} \leqslant \dfrac{1}{\beta}$,

(4.16) $$|u-\pi_u|^2_{H^1(\Omega)} \leqslant \sum_{K \in T_h} C^2 \frac{h_K^4}{\varrho_K^2} |u|^2_{H^2(K)} \leqslant \sum_{K \in T_h} \frac{C^2 h_K^2}{\beta^2} |u|^2_{H^2(K)}$$

$$\leqslant \frac{C^2 h^2}{\beta^2} |u|^2_{H^2(\Omega)}$$

so that

(4.17) $$|u-\pi_h u|_{H^1(\Omega)} \leqslant \frac{Ch}{\beta} |u|_{H^2(\Omega)} = Ch|u|_{H^2(\Omega)},$$

if the constant β is included in the constant C, and

(4.18) $$\|u-\pi_h u\|_{L_2(\Omega)} \leqslant Ch^2 |u|_{H^2(\Omega)}.$$

4.3 Interpolation with polynomials of higher degree

The estimates (4.17) and (4.18) are typical examples of estimates for the interpolation error $u-\pi_h u$, in this case for interpolation with piecewise linear functions. If we work with piecewise polynomials of degree $r \geqslant 1$ on triangulations T_h satisfying (4.1), we have in the typical case the following estimates:

(4.19) $$\|u-\pi_h u\|_{L_2(\Omega)} \leqslant Ch^{r+1} |u|_{H^{r+1}(\Omega)},$$

(4.20) $$|u-\pi_h u|_{H^1(\Omega)} \leqslant Ch^r |u|_{H^{r+1}(\Omega)},$$

where the constant β is absorbed in the constant C in (4.20). If $V_h \subset H^2(\Omega)$, then we also have

(4.21) $$|u-\pi_h u|_{H^2(\Omega)} \leqslant Ch^{r-1} |u|_{H^{r+1}(\Omega)}.$$

Note that for each derivative of the error $u-\pi_h u$, the power of h on the right hand side drops by one. Note that the constant C in (4.19)–(4.21) only depends on the constant β in (4.1) and the degree r, but not on the mesh parameter h or the function u.

Remark 4.1 If u does not have the regularity required in (4.19) or (4.20), we get the corresponding reduction in the power of h: For $1 \leqslant s \leqslant r+1$, we have

(4.22) $\quad \|u-\pi_h u\|_{L_2(\Omega)} \leqslant Ch^s |u|_{H^s(\Omega)},$

(4.23) $\quad \|u-\pi_h u\|_{H^1(\Omega)} \leqslant Ch^{s-1} |u|_{H^s(\Omega)}.$ □

Example 4.1 Let $\{T_h\}$ be a family of triangulations $T_h=\{K\}$ of $\Omega \subset R^2$ satisfying (4.1) and let $V_h=\{v \in C^0(\bar{\Omega}): v|_K \in P_2(K), \forall K \in T_h\}$. For the finite element of Example 3.2 we may for $v \in C^0(\bar{\Omega})$ define the interpolant $\pi_h v \in V_h$ by

$$\pi_h v = v \text{ at the nodes of } T_h,$$
$$\pi_h v = v \text{ at the midpoints of the sides of } T_h.$$

In this case (4.19) and (4.20) hold with r=2. □

Example 4.2 With $T_h=\{K\}$ as in Example 4.1 define $V_h=\{v \in C^1(\bar{\Omega}): v|_K \in P_5(K), \forall K \in T_h\}$ and for $v \in C^2(\bar{\Omega})$ specify the interpolant $\pi_h v \in V_h$ by

$$D^\alpha \pi_h v = D^\alpha v \text{ at the nodes of } T_h, \ |\alpha| \leqslant 2,$$

$$\frac{\partial}{\partial n} \pi_h v = \frac{\partial v}{\partial n} \text{ at the midpoints of each side S of } T_h,$$

where $\frac{\partial}{\partial n}$ denotes differentiation in the normal direction to S. In this case (4.19)–(4.21) hold with r=5. □

4.4 Error estimates for FEM for elliptic problems

Recalling again the typical abstract error estimate for an elliptic problem

$$\|u-u_h\|_V \leqslant C\|u-v\|_V \quad \forall v \in V_h,$$

and choosing here $v=\pi_h u$ with $\pi_h u \in V_h$ and interpolant of u, we have

(4.24) $\quad \|u-u_h\|_V \leqslant C\|u-\pi_h u\|_V \quad \forall v \in V_h.$

Using estimates for the interpolation error $||u-\pi_h u||_V$ we then obtain estimates for the finite element error $||u-u_h||_V$. Using the interpolation estimates of Sections 4.2 and 4.3 we have for example the following error estimates:

Example 4.3 With $V=H_0^1(\Omega)$ and (cf Examples 3.1–3.3)

$$V_h=\{v\in V: v|_K\in P_r(K), \forall K\in T_h\}, r=1, 2, 3,$$

we obtain from (4.20) and (4.24)

$$||u-u_h||_{H^1(\Omega)}\leq Ch^r|u|_{H^{r+1}(\Omega)}$$

for the finite element method for the Dirichlet problem (1.16). We obtain a similar result for the Neumann problem (1.36). □

Example 4.4 With V_h as in Example 4.2 we have for the biharmonic problem of Example 2.5 the following estimate

$$||u-u_h||_{H^2(\Omega)}\leq Ch^4|u|_{H^6(\Omega)}. \quad \square$$

Remark 4.2 It is possible to prove analogues of (4.24) in norms other than that given by the space V. For example one can prove for the finite element method of Section 1.4 that (see [RS])

$$||\nabla u-\nabla u_h||_{L_\infty(\Omega)}\leq C||\nabla u-\nabla \pi_h u||_{L_\infty(\Omega)},$$

which together with Theorem 4.1 gives

$$(4.25) \qquad ||\nabla u-\nabla u_h||_{L_\infty(\Omega)}\leq C \max_K [h_K \max_{|\alpha|=2} ||D^\alpha u||_{L_\infty(K)}]. \quad \square$$

4.5 On the regularity of the exact solution

We have seen that the regularity of the exact solution u is involved in estimating the error $||u-u_h||_V$ in the finite element method. Let us now give a typical result that shows how the regularity of the exact solution u depends on the regularity of the given data. Let us then consider the Poisson equation:

$$(4.26) \qquad \begin{aligned} -\Delta u&=f \quad \text{in } \Omega,\\ u&=0 \quad \text{on } \Gamma, \end{aligned}$$

where Ω is bounded domain in R^2 with boundary Γ and f is a given function. Let us first assume that Γ is smooth, i e, Γ is a smooth curve in particular without corners or cups. In this case there is for $s=0, 1, \ldots$, a constant C independent of f such that

$$(4.27) \qquad \|u\|_{H^{s+2}(\Omega)} \leq C\|f\|_{H^s(\Omega)},$$

i e, if $f \in H^s(\Omega)$ then $u \in H^{s+2}(\Omega)$, or loosely speaking, we "gain two derivatives" in (4.26).

If Γ is not smooth, then (4.27) may not hold, not even for $s=0$. If Γ has a corner, then the solution u or derivatives of u will in general have singularities at the corner even if f is very smooth ($f \in H^s(\Omega)$ for s large). More precisely, the solution u of (4.26) with f smooth basically has the following form close to a corner with angle ω (cf Problem 4.6):

$$(4.28) \qquad u(r,\theta)=r^{\gamma}\alpha(\theta)+\beta(r,\theta), \quad \gamma=\frac{\pi}{\omega},$$

where α and β are smooth functions (here we use polar coordinates (r, θ) with the pole at the corner). It is easy to see that if $\omega>\pi$ then a function u of the form (4.28) does not belong to $H^2(\Omega)$ if $\alpha \neq 0$. On the other hand, one can show that (4.27) holds with $s=0$ if Ω is a convex polygonal domain (in which case the corner angles satisfy $\omega<\pi$).

For the biharmonic problem (2.22) we have if the boundary Γ is smooth, for $s=0, 1, \ldots$,

$$\|u\|_{H^{s+4}(\Omega)} \leq C\|f\|_{H^s(\Omega)}.$$

If Γ has corners there are results analogous to those just stated for the Poisson equation (4.26).

Example 4.5 For a solution u of the form (4.28) we have formally that $u \in H^s(\Omega) \Leftrightarrow$ derivatives $D^s u$ of order s belong to $L_2(\Omega) \Leftrightarrow$

$$\int_{\Omega} |D^s u|^2 dx \sim C \int_0^R [r^{\gamma-s}]^2 r dr < \infty.$$

Hence $u \in H^s(\Omega)$ if and only if $s<\gamma+1$. By Remark 4.1 we thus have for the standard finite element method of Section 1.4 for the Poisson equation in a polygonal domain that for any $\varepsilon>0$

$$(4.29) \qquad \|u-u_h\|_{H^1(\Omega)} \leq Ch^{\gamma-\varepsilon}\|u\|_{H^{\gamma+1-\varepsilon}(\Omega)}=Ch^{\gamma-\varepsilon},$$

where $\gamma=\pi/\omega$ and ω is the maximal angle of a corner of Γ. For example if $\gamma={}^2/_3$, which corresponds to a concave corner of angle $3\pi/2$, then

$$\|u-u_h\|_{H^1(\Omega)} \leq Ch^{\frac{2}{3}-\varepsilon}.$$

We see that in this case we do not obtain the full rate of convergence which is $0(h)$. □

4.6 Adaptive methods

If the exact solution u has e g a corner singularity, then it is natural to refine the triangulation close to the corner to increase the accuracy. Recalling that for the method of Section 1.4 (cf (4.16))

(4.30) $$|u-u_h|_{H^1(\Omega)} \leq |u-\pi_h u|_{H^1(\Omega)} \leq C[\Sigma(h_K|u|_{H^2(K)})^2]^{1/2},$$

it is clear that we somehow would like to balance the size of h_K with that of $|u|_{H^2(K)}$ and in particular choose h_K small where $|u|_{H^2(K)}$ is large. If u has the form (4.28) with $0<\gamma<1$, then one possible refinement is given by (cf Problem 4.4.)

(4.31) $$h_K = Chd_K^{1-\gamma},$$

if $h_K \leq d_K$, where d_K is the distance from K to the corner and h is the mesh size away from the corner. With this refinement we have, disregarding the ε in (4.29),

(4.32) $$|u-u_h|_{H^1(\Omega)} \leq Ch.$$

Notice that the total number of elements with a refinement of the form (4.31) is of the order $0(h^{-2})$, i.e., the same as with a uniform mesh of size h. Thus, in this case the refinement does not increase the total number of unknowns significantly but significantly increases the precision (from (4.29) to (4.32)).

In general the nature of the exact solution u is not known beforehand and then it is not clear how to locally refine the finite element mesh. Recently methods for automatic mesh refinement, so-called *adaptive methods,* have been developed which do not require the user to supply information on the smoothness of the exact solution. In these methods this information is instead obtained through a sequence of computed solutions on successively refined meshes.

To very briefly describe some of the basic ideas underlying adaptive methods, suppose $\delta>0$ is a given *tolerance* and suppose we want to obtain a finite element approximation u_h such that

(4.33) $$|u-u_h|_{H^1(\Omega)} \lesssim \delta.$$

Relying on the error estimate (4.30) we see that (4.33) will be satisfied if the corresponding finite element mesh $T_h=\{K\}$ is chosen so that

$$(4.34) \qquad \sum_{K\in T_h} (h_K|u|_{H^2(K)})^2 \sim (\frac{\delta}{C})^2.$$

To determine a mesh satisfying (4.34) we may proceed as follows: Choose a first mesh $\bar{T}_h=\{\bar{K}\}$ and compute a corresponding finite element solution $\bar{u}_h$. Using $\bar{u}_h$ compute approximations to $|u|_{H^2(\bar{K})}$ denoted by $|\bar{u}_h|_{H^2(\bar{K})}$ for $\bar{K}\in\bar{T}_h$. The quantity $|\bar{u}_h|_{H^2(\bar{K})}$ may be obtained using difference quotients based on the values of $\nabla\bar{u}_h$ at the centers of gravity of $\bar{K}$ and neighbouring triangles in $\bar{T}_h$. Next, construct a new mesh $T_h=\{K\}$ by subdividing into four equal triangles each $\bar{K}\in\bar{T}_h$ for which

$$(h_{\bar{K}}|\bar{u}_h|_{H^2(\bar{K})})^2 > \frac{\delta^2}{\bar{N}C^2},$$

where $\bar{N}$ is the number of triangles in $\bar{T}_h$. Next, compute the finite element solution u_h on the new mesh T_h and repeat the process until

$$(4.35) \qquad \sum_{K\in T_h} (h_K|u_h|_{H^2(K)})^2 \leq (\frac{\delta}{C})^2.$$

Note that by the construction if follows (if δ is small enough) that for the final mesh T_h satisfying (4.35), all the terms in the sum will be approximately equal. Note also that after refinement of certain triangles, the resulting mesh is completed into a triangulation as in Fig 1.15.

It is also possible to control the error in other norms than the $H^1(\Omega)$-norm used in (4.33), for instance we may want to control the gradient error in the maximum norm. In this case we base the adaptive method on the error estimate (4.25) and seek to find a mesh $T_h=\{K\}$ such that

$$(4.36) \qquad Ch_K \max_{|\alpha|=2} \|D^\alpha u_h\|_{L_\infty(K)} \sim \delta \quad \forall K\in T_h,$$

where as above $\|D^\alpha u_h\|_{L_\infty(K)}$ is a computed approximation of $\|D^\alpha u\|_{L_\infty(K)}$. Again the final mesh satisfying (4.36) is constructed through a sequence of successively refined meshes where triangles K for which the left hand side of (4.36) is larger than δ are refined. In Fig. 4.2 we give the sequence of meshes (with a zoom at the origin for the final mesh) obtained by applying an adaptive method of this form with $\delta=0.1$ and $C=1$ to the problem

$$\Delta u=0 \text{ in } \Omega,$$
$$u=u_0 \text{ on } \Gamma,$$

where $\Omega=\{x=r(\cos\theta, \sin\theta): 0<r<1, 0<\theta<3\pi/4\}$ with exact solution $u(r, \theta)=r^{\gamma}\sin(\gamma\theta)$, $\gamma=4/3$. In Fig. 4.3 we give the actual gradient error $|\nabla e(x)|$ as a function of the distance $|x|$ to the origin along the radius $\theta=\pi/2$. We observe that the gradient error is roughly equal to the tolerance and thus we see that the adaptive method is able to find a good mesh in this case. This example is taken from [EJ2], where theoretical and computational results for adaptive methods of the indicated type are given, see also [E]. The problem of computationally estimating the constant C in (4.32) and (4.34) is discussed in [EJ2].

For adaptive methods for parabolic problems we refer to Section 8.4.4. For another approach to adaptivity, see [BR], [BM].

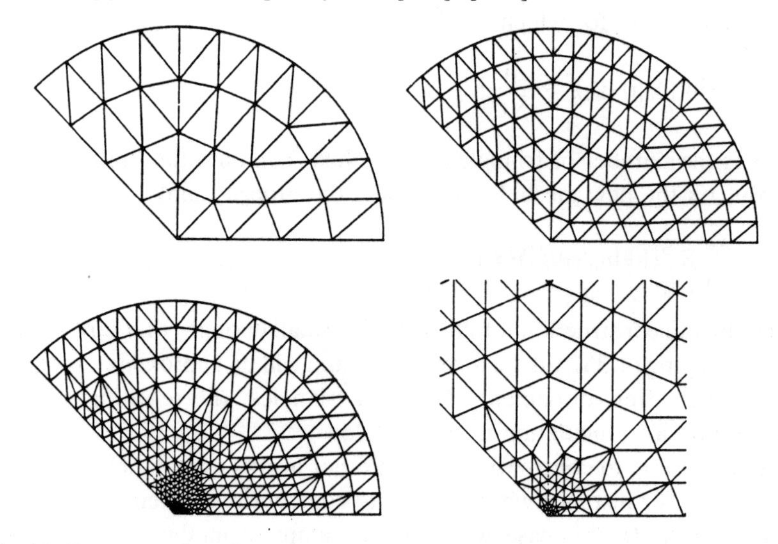

Fig 4.2 Sequence of meshes obtained by adaptive FEM

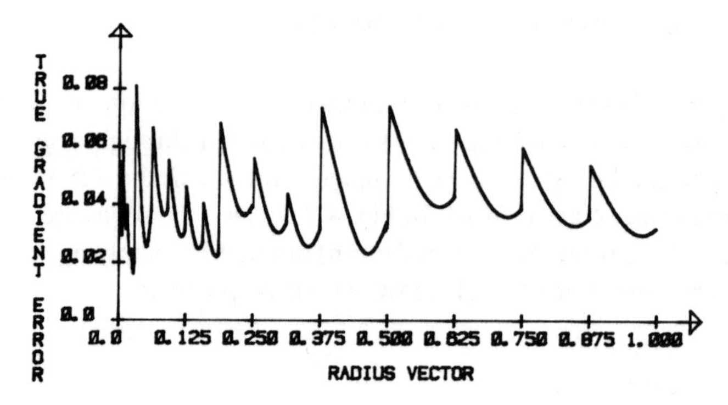

Fig 4.3 Gradient error

4.7 An error estimate in the $L_2(\Omega)$-norm

We have seen that if we apply the finite element method with the space $V_h=\{v\in H_0^1(\Omega): v|_K\in P_1(K), \forall K\in T_h\}$ to the Poisson equation (1.16) with Ω a polygonal domain, then we have the following estimate for the error $u-u_h$ in the $H^1(\Omega)$-norm:

$$\|u-u_h\|_{H^1(\Omega)}\leq Ch|u|_{H^2(\Omega)}. \tag{4.37}$$

This trivially gives the following $L_2(\Omega)$-estimate:

$$\|u-u_h\|_{L_2(\Omega)}\leq Ch|u|_{H^2(\Omega)}. \tag{4.38}$$

On the other hand by (4.18) the interpolation error, $u-\pi_h u$, satisfies the second order estimate:

$$\|u-\pi_h u\|_{L_2(\Omega)}\leq Ch^2|u|_{H^2(\Omega)}.$$

We shall now prove that we have a similar estimate for $\|u-u_h\|_{L_2(\Omega)}$ so that this quantity in fact converges at the optimal rate. We shall then assume that the polygonal domain Ω is convex (if Ω has a smooth boundary, then convexity is not required).

Theorem 4.3. If Ω is a convex polygonal domain and u_h is the finite element solution of the Poisson equation (1.16) with piecewise linear functions, i e u_h satisfies (1.20), then there is a constant independent of u and h such that

$$\|u-u_h\|_{L_2(\Omega)}\leq Ch^2|u|_{H^2(\Omega)}.$$

Proof. Subtracting (1.19) and (1.20) we obtain the error equation

$$a(e,v)=0 \quad \forall v\in V_h, \tag{4.39}$$

where $e=u-u_h$ and the notation of (1.19) is used. We shall now estimate $(e,e)=\|e\|^2_{L_2(\Omega)}$ using a so-called *duality* argument which is often used in finite element analysis (see also Chapter 8). Let φ be the solution of the following auxiliary dual problem:

$$\begin{aligned} -\Delta\varphi&=e \quad \text{in } \Omega,\\ \varphi&=0 \quad \text{on } \Gamma. \end{aligned}$$

Since Ω is convex we have from (4.27) with $s=0$,

$$\|\varphi\|_{H^2(\Omega)}\leq C\|e\|_{L_2(\Omega)}, \tag{4.40}$$

where the constant C does not depend on e. Using Green's formula and the fact that e=0 on Γ,

$$(e, e)=-(e, \Delta\varphi)=a(e, \varphi)=a(e, \varphi-\pi_h\varphi),$$

where the last inequality follows from (4.39) since $\pi_h\varphi\in V_h$ so that $a(e, \pi_h\varphi)=0$. Applying now the interpolation estimate (4.18) to φ and using also (4.40), we find

$$\|e\|^2_{L_2(\Omega)}\leq\|e\|_{H^1(\Omega)}\|\varphi-\pi_h\varphi\|_{H^1(\Omega)}\leq C\|e\|_{H^1(\Omega)}h|\varphi|_{H^2(\Omega)}$$
$$\leq Ch\|e\|_{H^1(\Omega)}\|e\|_{L_2(\Omega)}.$$

Dividing by $\|e\|_{L_2(\Omega)}$ and recalling (4.37) we finally get

$$\|e\|_{L_2(\Omega)}\leq Ch\|e\|_{H^1(\Omega)}\leq Ch^2|u|_{H^2(\Omega)}$$

and the proof is complete. □

Remark 4.3 The basic stability inequality (2.6) for (4.26) states that

$$\|u\|_{H^1(\Omega)}\leq\frac{\Lambda}{\alpha}, \tag{4.41}$$

where Λ is any constant such that

$$|L(v)|=|(f, v)|\leq\Lambda\|v\|_{H^1(\Omega)}\qquad\forall v\in H_0^1(\Omega).$$

The smallest possible choice of Λ is given by

$$\Lambda=\sup_{\substack{v\in H_0^1(\Omega)\\ v\neq0}}\frac{|(f, v)|}{\|v\|_{H^1(\Omega)}}. \tag{4.42}$$

Clearly the quantity Λ defined by (4.42) measures the size of f in a certain sense and in fact we may define a norm $\|\cdot\|_{H^{-1}(\Omega)}$ by

$$\|f\|_{H^{-1}(\Omega)}=\sup_{\substack{v\in H_0^1(\Omega)\\ v\neq0}}\frac{|(f, v)|}{\|v\|_{H^1(\Omega)}}. \tag{4.43}$$

This is the norm in the so-called *dual space* $H^{-1}(\Omega)$ of $H_0^1(\Omega)$. Note that

$$\|f\|_{L_2(\Omega)}=\sup_{\substack{v\in L_2(\Omega)\\ v\neq0}}\frac{|(f, v)|}{\|v\|_{L_2(\Omega)}},$$

and since we take sup over a larger set, we clearly have that $||\cdot||_{L_2(\Omega)}$ is a stronger norm than $||\cdot||_{H^{-1}(\Omega)}$, i e,

$$||f||_{H^{-1}(\Omega)} \leq ||f||_{L_2(\Omega)}.$$

By (4.42) and (4.43) it follows that the basic stability inequality (4.41) may be written as

$$||u||_{H^1(\Omega)} \leq \frac{1}{\alpha}||f||_{H^{-1}(\Omega)},$$

which formally corresponds to (4.27) with $s=-1$. □

Problems

4.1 Let $I=[0, h]$ and let $\pi v \in P_1(I)$ be the linear interpolant that agrees with $v \in C^0(I)$ at the end points of I. Using the technique of the proof of Theorem 4.1 prove estimates for $||v-\pi v||_{L_\infty(I)}$ and $||v'-(\pi v)'||_{L_\infty(I)}$, cf (1.12) and (1.13).

4.2 Prove (4.14).

4.3 Estimate the error $||u-u_h||_{H^2(I)}$ for Problem 1.5 and Example 2.4.

4.4 Prove that the total number of elements with a corner refinement of the form (4.31) is $0(h^{-2})$.

4.5 Determine a suitable refinement in case the exact solution has a singularity of the form (4.28) with $1<\gamma<2$ and we want to control $||\nabla u-\nabla u^h||_{L_\infty(\Omega)}$ via the estimate (4.25), cf [EJ2].

4.6 Using polar coordinates (r, θ), let $\Omega=\{(r, \theta): 0<r<1, 0<\theta<\omega\}$ be a pie-shaped domain of angle ω. Prove that the function $u(r, \theta) = r^\gamma \sin(\gamma\theta)$, $\gamma=\frac{\pi}{\omega}$, satisfies: $\Delta u=0$ in Ω, $u=0$ on the straight parts of the boundary of Ω.

4.7 Prove, by modifying the proof of Theorem 4.3, the following L_2-estimate for the standard finite element method of Section 1.4 for Poisson's equation on an L-shaped domain (cf Example 4.5):

$$||u-u_h||_{L_2(\Omega)} \leq Ch^{4/3-\varepsilon}.$$

4.8 Let V_h be a finite element space on a triangulation T_h of the domain $\Omega \subset R^d$ satisfying (4.19). Given $u \in L_2(\Omega)$ let $u_h \in V_h$ be the $L_2(\Omega)$-*projection of* u *onto* V_h, i e,

(4.44) $(u_h, v)=(u, v) \qquad \forall v\in V_h,$

where $(.,.)$ is the scalar product in $L_2(\Omega)$. Prove the error estimate

$$\|u-u_h\|_{L_2(\Omega)}\leqslant\inf_{v\in V_h}\|u-v\|_{L_2(\Omega)}\leqslant Ch^{r+1}|u|_{H^{r+1}(\Omega)},$$

and that

$$\|u_h\|_{L_2(\Omega)}\leqslant\|u\|_{L_2(\Omega)}.$$

5. Some applications to elliptic problems

This chapter presents applications of the finite element method to some basic problems in continuum mechanics of elliptic type. We first give suitable variational formulations of the continuous problems.

5.1 The elasticity problem

Consider a homogenous isotropic elastic body *B* occupying the bounded domain $\Omega \subset R^3$ with boundary Γ decomposed into two parts Γ_1 and Γ_2 with the area of Γ_2 being positive. Let *B* be acted upon by a volume load $f=(f_1, f_2, f_3)$ and a boundary load $g=(g_1, g_2, g_3)$ on Γ_1, where the f_i and g_i are the components in the x_i-direction. Further, let us assume that *B* is fixed along Γ_2 (see Fig 5.1).

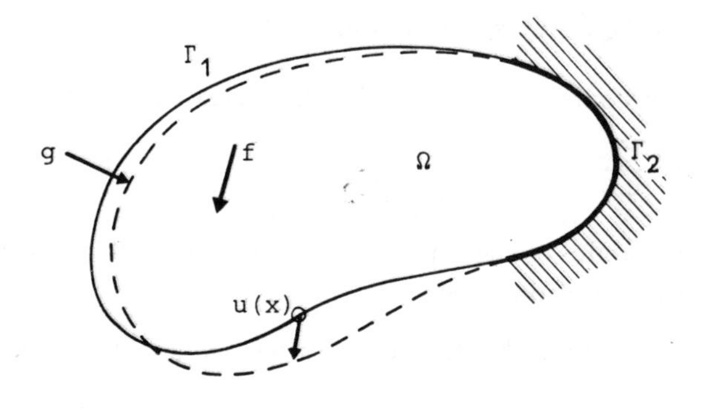

Fig 5.1

We want to determine the *displacement* $u=(u_1, u_2, u_3)$ and the symmetric *stress tensor* $\sigma=(\sigma_{ij})$, $\sigma_{ij}=\sigma_{ij}$, i, j=1, 2, 3, under the loads f and g. Here u_i is the displacement in the x_i-direction, σ_{ii} is the *normal stress* in the x_i-direction, and the σ_{ij}, $i\neq j$, are the *shear stresses*. Further, $\varepsilon(u)=(\varepsilon_{ij}(u))$, where

$$\varepsilon_{ij}(u)=\frac{1}{2}\left(\frac{\partial u_i}{\partial x_j}+\frac{\partial u_j}{\partial x_i}\right), \qquad i, j=1, 2, 3,$$

is the *deformation* (tensor) associate with the displacement u. Assuming that *B* is linearly elastic and that the displacements are small, we have the following relation between stresses and deformations, or *constitutive relation (Hooke's law):*

$$\sigma_{ij}=\lambda \operatorname{div} u\, \delta_{ij}+\mu\varepsilon_{ij}(u), \tag{5.1a}$$

where λ and μ are positive constants,

$$\operatorname{div} u=\sum_{i=1}^{3}\frac{\partial u_i}{\partial x_i},$$

$$\delta_{ij}=\begin{cases}1 \text{ if } i=j,\\ 0 \text{ if } i\neq j.\end{cases}$$

We also have the *equilibrium equations*

$$-\sum_{j=1}^{3}\frac{\partial \sigma_{ij}}{\partial x_j}=f_i \quad \text{in } \Omega,\ i=1, 2, 3, \tag{5.1b}$$

together with the boundary conditions

$$u=0 \text{ on } \Gamma_2, \tag{5.1c}$$

$$\sum_{j=1}^{3}\sigma_{ij}n_j=g_i \qquad \text{on } \Gamma_1,\ i=1, 2, 3, \tag{5.1d}$$

where $n=(n_i)$ is the outward unit normal to Γ.

Remark The constants μ and λ in (5.1a) can be expressed as

$$\mu=\frac{E}{1+\nu}, \quad \lambda=\frac{E\nu}{(1+\nu)\,(1-2\nu)},$$

where E is the modulus of elasticity (Young modulus) and ν is the contraction ratio (Poisson ratio) of the elastic material of *B*. □

In the remainder of this chapter the following notation for partial derivatives will be used

$$v_{,j}=\frac{\partial v}{\partial x_j}, \; j=1, 2, 3.$$

We shall also use the summation convention that repeated indices indicate summation from 1 to 3. With this convention we may write the equilibrium equations (5.1b) as follows

$$-\sigma_{ij,j}=f_i \quad \text{in } \Omega,\ i=1, 2, 3.$$

We will now give a variational formulation of the elasticity problem (5.1). Let us then first note the following Green's formula:

$$(5.2) \qquad \int_\Omega \sigma_{ij}\varepsilon_{ij}(v)dx=\int_\Gamma \sigma_{ij}n_iv_jds-\int_\Omega \sigma_{ij,j}\, v_idx,$$

where the summation convention is applied in all terms, i e, we sum over i and j from 1 to 3. To show (5.2) observe that since $\sigma_{ij}=\sigma_{ji}$ and $\varepsilon_{ij}(v)=\frac{1}{2}(v_{i,j}+v_{j,i})$, we have

$$\sigma_{ij}\varepsilon_{ij}(v)=\frac{1}{2}(\sigma_{ij}v_{i,j}+\sigma_{ji}v_{j,i})=\frac{1}{2}(\sigma_{ij}v_{i,j}+\sigma_{ij}v_{i,j})=\sigma_{ij}v_{i,j}.$$

Hence, by Green's formula (1.17) we get

$$\int_\Omega\sigma_{ij}\varepsilon_{ij}(v)dx=\int_\Omega\sigma_{ij}v_{i,j}dx=\int_\Gamma\sigma_{ij}n_jv_ids-\int_\Omega\sigma_{ij,j}v_idx,$$

which proves (5.2). Let us next choose a test function $v=(v_1, v_2, v_3)\in[H^1(\Omega)]^3$ (i e each component $v_i\in H^1(\Omega)$) such that $v=0$ on Γ_2, multiply (5.1b) by v_i, sum over i from 1 to 3 and integrate over Ω. By Green's formula (5.2), we then have

$$\int_\Omega f_iv_idx=-\int_\Omega\sigma_{ij,j}v_idx=\int_\Omega\sigma_{ij}\varepsilon_{ij}(v)dx-\int_{\Gamma_1}\sigma_{ij}n_jv_ids,$$

where the boundary integral over Γ_2 vanishes since $v=0$ on Γ_2. Using also (5.1d) we thus have

$$\int_\Omega\sigma_{ij}\varepsilon_{ij}(v)dx=\int_\Omega f_iv_idx+\int_{\Gamma_1}g_iv_ids.$$

Finally, we eliminate σ_{ij} by using (5.1a) to get

$$\int_\Omega[\lambda \text{ div } u \text{ div } v+\mu\varepsilon_{ij}(u)\varepsilon_{ij}(v)]dx=\int_\Omega f_iv_idx+\int_{\Gamma_1}g_iv_ids,$$

since

$$\text{div } u\ \delta_{ij}\varepsilon_{ij}(v)=\text{div } u \text{ div } v.$$

We are thus led to the following variational formulation of the elasticity problem (5.1): Find $u\in V$ such that

$$(5.3) \qquad a(u, v)=L(v) \qquad \forall v\in V,$$

where

$$a(u, v)=\int_\Omega[\lambda \text{ div } u \text{ div } v+\mu\varepsilon_{ij}(u)\varepsilon_{ij}(v)]dx,\ L(v)=\int_\Omega f_iv_idx+\int_{\Gamma_1}g_iv_ids,$$

$$V=\{v\in[H^1(\Omega)]^3\colon v=0 \text{ on } \Gamma_2\}.$$

Let us now check if the assumptions (i)–(iv) of Section 2.1 are satisfied in this case. We can routinely verify (i), (ii) and (iv) and thus it only remains to prove the V-ellipticity, i e, to prove that there is a positive constant α such that

$$(5.4) \qquad a(v, v) \geqslant \alpha \|v\|_V^2 \qquad \forall v \in V,$$

where

$$\|v\|_V = \|v\|_{H^1(\Omega)} = \left(\sum_{i=1}^{3} \|v_i\|_{H^1(\Omega)}^2\right)^{1/2}.$$

This inequality follows directly from *Korn's inequality:* There is a positive constant c such that

$$(5.5) \qquad \int_\Omega \varepsilon_{ij}(v)\varepsilon_{ij}(v)dx \geqslant c\|v\|_V^2 = c(|v|_{H^1(\Omega)}^2 + \|v\|_{L_2(\Omega)}^2).$$

We notice in particular that (5.5) amounts to proving that the L_2-norm of any partial derivative $v_{i,j}$ can be estimated by the L_2-norms of the deformations $\varepsilon_{ij}(v)$ involving only certain combinations of the $v_{i,j}$. Since (5.5) involves also the L_2-norm on the right hand side, we need Γ_2 to have positive measure (cf Example 2.7). For a proof of Korn's inequality we refer to [Ni] (the proof is easy in the case $\Gamma_2 = \Gamma$, cf Problem 5.2).

Now we are able to formulate a finite element method for our elasticity problem. Let then $T_h = \{K\}$ be a "triangulation" of Ω into tetrahedrons K as described in Example 3.6 and define

$$V_h = \{v \in V: v|_K \in [P_1(K)]^3, \forall K \in T_h\}.$$

Each component v_i of a function $v \in V_h$ is thus a piecewise linear function vanishing on Γ_2. We now formulate the following finite element method for (5.1): Find $u_h \in V_h$ such that

$$a(u_h, v) = L(v) \qquad \forall v \in V_h.$$

According to the general theory of Chapter 2 this problem has a unique solution and by the interpolation results of Chapter 4 we have the following error estimate:

$$\|u - u_h\|_{H^1(\Omega)} \leqslant Ch|u|_{H^2(\Omega)}.$$

Problems

5.1 Consider the elasticity problem (5.1) in a three-dimensional domain $\hat{\Omega} = \Omega \times (-\varepsilon, \varepsilon)$ with $\Omega \subset R^2$, ε small and $F_3 = f_3 = 0$. This corresponds to a thin elastic plate with middle surface Ω subject to in-plane loads only (no transversal loads). Assuming "plane stresses" (ie, $\sigma_{i3}=0$, $i=1, 2, 3$) prove that in this case (5.1) is reduced to:

$$\sigma_{ij}=\bar{\lambda}(\varepsilon_{11}(u)+\varepsilon_{22}(u))\delta_{ij}+\bar{\mu}\ \varepsilon_{ij}(u), \qquad i, j=1, 2,$$

$$-\sum_{j=1}^{2}\frac{\partial\sigma_{ij}}{\partial x_j}=f_i \text{ in } \Omega, \quad i, j=1, 2,$$

$$u_1=u_2=0 \qquad \text{on } \Gamma_1, \quad i=1, 2,$$

$$\sum_{j=1}^{2}\sigma_{ij}n_j=F_i \qquad \text{on } \Gamma_2, \quad i=1, 2,$$

where $\bar{\lambda}=\dfrac{E\nu}{1-\nu^2}, \bar{\mu}=\dfrac{E}{1+\nu}$, Γ_1 and Γ_2 is a decomposition of the boundary of Ω and f_i and F_i are given forces. Give a variational formulation of this problem and formulate a corresponding finite element method. Determine the stiffness matrix in a problem with simple triangulation and piecewise linear displacements. In Fig 5.2 below we give the computed displacements using bilinear elements on the indicated triangulation for the above problem corresponding to a thin plate fixed at both ends and subjects to a distributed load as indicated. The Young modulus E is here different in the upper and lower halfs of the plate denoted by I and II, with E being larger in II.

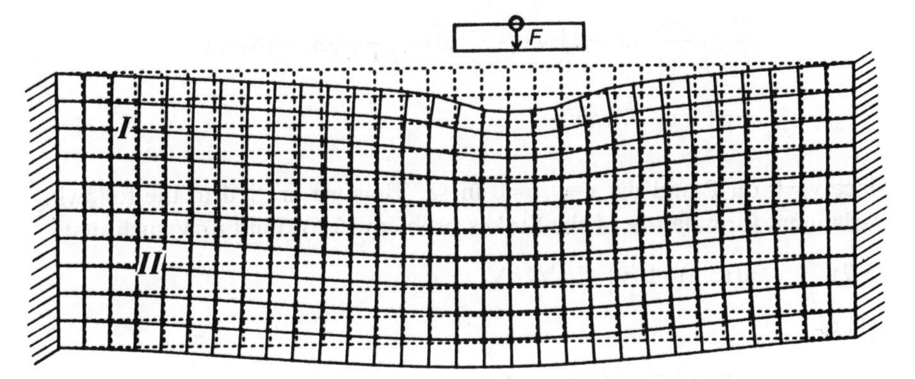

Fig 5.2

5.2. Prove Korn's inequality in the case $\Gamma_2=\Gamma$. Hint: Prove that

$$\int_\Omega v_{i,j}v_{j,i}dx=\int_\Omega v_{i,i}v_{j,j}dx.$$

5.2 Stokes problem

The stationary Stokes equations for an incompressible Newtonian fluid with viscosity μ, enclosed in the domain $\Omega \subset R^3$, and acted upon by the volume load f, read in the notation of Section 5.1:

$$\begin{aligned}
\sigma_{ij} &= 2\mu\varepsilon_{ij}(u) - p\,\delta_{ij} && \text{in } \Omega, \quad i, j=1, 2, 3,\\
-\sigma_{ij,j} &= f_i && \text{in } \Omega, \quad i=1, 2, 3,\\
\operatorname{div} u &= 0 && \text{in } \Omega,\\
u &= 0 && \text{on } \Gamma,
\end{aligned}$$

where $\sigma=(\sigma_{ij})$ is the stress, p the pressure and $u=(u_i)$ the velocity. Eliminating σ_{ij} we obtain the following equivalent formulation:

$$-\mu\Delta u_i + p_{,i} = f_i \quad \text{in } \Omega, \quad i=1, 2, 3, \tag{5.6a}$$

$$\operatorname{div} u = 0 \quad \text{in } \Omega, \tag{5.6b}$$

$$u_i = 0 \quad \text{on } \Gamma, \quad i=1, 2, 3. \tag{5.6c}$$

We now seek a variational formulation of (5.6). Let $v \in [H_0^1(\Omega)]^3$ be a test function satisfying the incompressibility condition div $v=0$ in Ω, multiply (5.6a) by v_i, integrate over Ω and use Green's formula. Then summing over i, we get

$$\begin{aligned}
\int_\Omega f_i v_i dx &= -\mu \int_\Omega \Delta u_i v_i dx + \int_\Omega p_{,i} v_i dx\\
&= -\int_\Gamma \frac{\partial u_i}{\partial n} v_i ds + \mu \int_\Omega \nabla u_i \cdot \nabla v_i dx + \int_\Gamma p n_i v_i ds - \int_\Omega p v_{i,i} dx\\
&= \mu \int_\Omega \nabla u_i \cdot \nabla v_i dx,
\end{aligned}$$

since $v_i=0$ on Γ and div $v \equiv v_{i,\,i}=0$ in Ω. Thus we are led to the following variational formulation of the Stokes problem (5.6): Find $u \in V$ such that

$$a(u, v) = L(v) \quad \forall v \in V, \tag{5.7}$$

where

$$a(v, w) = \mu \int_\Omega \nabla v_i \cdot \nabla w_i dx,$$

$$L(v) = \int_\Omega f_i v_i dx,$$

$$V = \{v \in [H_0^1(\Omega)]^3 : \operatorname{div} v = 0 \quad \text{in } \Omega\}.$$

We can easily check that the conditions (i)–(iv) of Section 2.1 are satisfied. Note that in the formulation (5.7) the pressure p has "disappeared", which comes from the fact that we are working with a space V for the velocities where the incompressibility condition $\operatorname{div} v=0$ is satisfied.

To formulate a finite element method for (5.6) based on the variational formulation (5.7) we need to construct a finite-dimensional subspace V_h of V. It turns out that this is not altogether easy since we have to satisfy the condition $\operatorname{div} v=0$ exactly. For simplicity, let us consider the analogue of (5.7) in two dimensions, in which case

$$V=\{v=(v_1, v_2)\in[H_0^1(\Omega)]^2: \operatorname{div} v\equiv \frac{\partial v_1}{\partial x_1}+\frac{\partial v_2}{\partial x_2}=0 \text{ in } \Omega\},$$

where $\Omega\subset R^2$. By a standard result in advanced calculus it follows that if Ω is simply connected, i e, if Ω does not contain any "holes", then $\operatorname{div} v=0$ in Ω if and only if

$$v=\operatorname{rot} \varphi\equiv\left(\frac{\partial\varphi}{\partial x_2}, -\frac{\partial\varphi}{\partial x_1}\right),$$

for some function φ. More precisely (cf Problem 5.2), one has

$$v\in V \Leftrightarrow v=\operatorname{rot} \varphi,\ \varphi\in H_0^2(\Omega). \tag{5.8}$$

The function φ is the *stream function* connected with the velocity field v.

Let now W_h be a finite-dimensional subspace of $H_0^2(\Omega)$ e g constructed using the C^1-element of Example 3.5 and define

$$V_h=\{v: v=\operatorname{rot} \varphi,\ \varphi\in W_h\}.$$

Then $V_h\subset V$ and formulating a finite element method in the usual way by replacing V by V_h in (5.7) we obtain a discrete solution u_h satisfying the following error estimate:

$$\|u-u_h\|_{H^1(\Omega)}\leq Ch^4|u|_{H^5(\Omega)}.$$

Chapter 11 gives other finite element methods (so-called mixed methods) for the two dimensional analogue of the Stokes problem (5.6) not requiring the velocity space V_h to satisfy the incompressibility condition exactly.

5.3 A plate problem

Consider a thin elastic plate *P* with middle surface given by the domain $\Omega \subset R^2$ with boundary Γ and acted upon by the transversal load f, see Fig 5.3.

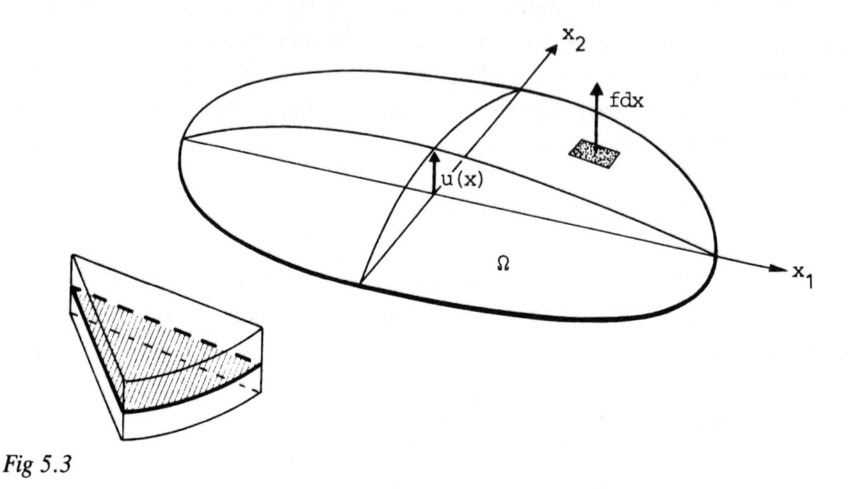

Fig 5.3

We seek the transversal *deflection* u together with the *moments* σ_{ij}, i, j=1, 2, under the load f. Here σ_{ii} is the *bending moment* in the x_i – direction and $\sigma_{12}=\sigma_{21}$ the *twisting moment.* Assuming small deflections and a linearly elastic material, we have the following constitutive relation (cf Hooke's law):

(5.9) $\quad \sigma_{ij}=\lambda\Delta u \, \delta_{ij}+\mu\varkappa_{ij}(u), \quad i=1, 2,$

where λ and μ are positive constants, and

$$\varkappa_{ij}(u)=u_{,ij}=\frac{\partial^2 u}{\partial x_i \partial x_j},$$

defines the *curvature* tensor. Further we have the following equilibrium equation:

(5.10) $\quad \sigma_{ij,ij}=f \quad \text{in } \Omega,$

where again the summation convention is used.

To define the boundary conditions let $n=(n_1, n_2)$ be the outward unit normal to Γ, $t\equiv(t_1, t_2)\equiv(n_2, -n_1)$ the tangent to Γ and define

$$\frac{\partial u}{\partial n}=u_{,j}n_j \qquad \text{(normal derivative)},$$

$$\frac{\partial u}{\partial t}=u_{,j}t_j \qquad \text{(tangential derivative)},$$

$$\sigma_{nn}=\sigma_{ij}n_in_j \qquad \text{(normal moment)},$$

$$\sigma_{nt}=\sigma_{ij}n_it_j \qquad \text{(twisting moment)},$$

$$R(\sigma)=\sigma_{ij,j}n_i+\frac{\partial \sigma_{nt}}{\partial t} \qquad \text{(transversal force)}.$$

Now let the boundary Γ be partitioned into three parts Γ_i, $i=1, 2, 3$, and consider the following boundary conditions:

$$u=\frac{\partial u}{\partial n}=0 \qquad \text{on } \Gamma_1 \text{ (clamped)}, \tag{5.11a}$$

$$u=\sigma_{nn}=0 \qquad \text{on } \Gamma_2 \text{ (simply supported)}, \tag{5.11b}$$

$$\sigma_{nn}=R(\sigma)=0 \qquad \text{on } \Gamma_3 \text{ (free boundary)}. \tag{5.11c}$$

Let us now give a variational formulation of the plate problem (5.9)–(5.11). Let $v\in H^2(\Omega)$ be a test function satisfying the essential boundary conditions

$$v=\frac{\partial n}{\partial n}=0 \qquad \text{on } \Gamma_1,$$

$$v=0 \qquad \text{on } \Gamma_2. \tag{5.12}$$

If we now multiply the equilibrium equation (5.10) by v and integrate over Ω, then repeated use of Green's formula gives

$$\begin{aligned}\int_\Omega fv\,dx&=\int_\Omega \sigma_{ij,ij}v\,dx=\int_\Gamma \sigma_{ij,j}n_iv\,ds-\int_\Omega \sigma_{ij,j}v_{,i}dx\\ &=\int_\Gamma \sigma_{ij,j}n_iv\,ds-\int_\Gamma \sigma_{ij}n_jv_{,i}ds+\int_\Omega \sigma_{ij}\varkappa_{ij}(v)dx.\end{aligned} \tag{5.13}$$

Since

$$v_{,i}=\frac{\partial v}{\partial n}n_i+\frac{\partial v}{\partial t}t_i,\ i=1, 2,$$

we have

$$\sigma_{ij}n_jv_{,i}=\sigma_{ij}n_jn_i\frac{\partial v}{\partial n}+\sigma_{ij}n_jt_i\frac{\partial v}{\partial t}=\sigma_{nn}\frac{\partial v}{\partial n}+\sigma_{nt}\frac{\partial v}{\partial t},$$

so that (5.13) can be written

$$\int_\Omega \sigma_{ij}\varkappa_{ij}(v)dx=\int_\Omega fv\,dx-\int_\Gamma \sigma_{ij,j}n_iv\,ds+\int_\Gamma \sigma_{nn}\frac{\partial v}{\partial n}ds+\int_\Gamma \sigma_{nt}\frac{\partial v}{\partial t}ds. \tag{5.14}$$

If the boundary Γ is smooth, partial integration along Γ gives

$$\int_\Gamma \sigma_{nt}\frac{\partial v}{\partial t}ds=-\int_\Gamma \frac{\partial \sigma_{nt}}{\partial t}v\ ds.$$

in which case (5.14) can be written

$$\int_\Omega \sigma_{ij}\varkappa_{ij}(v)dx=\int_\Omega fv\ dx+\int_\Gamma \sigma_{nn}\frac{\partial v}{\partial n}ds-\int_\Gamma R(\sigma)v\ ds.$$

If we now use the boundary conditions (5.11) and (5.12), we see that the boundary integrals disappear and on eliminating σ_{ij} also, by using (5.9), we finally get

$$\int_\Omega [\lambda\Delta u\Delta v+\mu\varkappa_{ij}(u)\varkappa_{ij}(v)]dx=\int_\Omega fv\,dx.$$

Thus we are led to the following variational formulation of the plate problem (5.9)–(5.11): Find $u\in V$ such that

$$a(u, v)=L(v) \qquad \forall v\in V,$$

where

$$a(u,v)=\int_\Omega [\lambda\Delta u\Delta v+\mu\varkappa_{ij}(u)\varkappa_{ij}(v)]dx,$$

$$L(v)=\int_\Omega fv\,dx,$$

$$V=\{v\in H^2(\Omega)\colon v=\frac{\partial v}{\partial n}=0 \quad \text{on } \Gamma_1,\ v=0 \text{ on } \Gamma_2\}.$$

We immediately see that the conditions (i), (ii) and (iv) of Section 2.1 are satisfied and it is possible to verify the V-ellipticity for example in the case when the length of Γ_1 is positive, i e, when P is clamped along a part the boundary (cf Problem 2.2).

We can now in a routine way formulate a finite element method for the plate problem using the C^1-element of Example 3.5. We leave the details to the reader.

Remark The constants λ and μ in (5.9) are given by

$$\lambda=\frac{Ea^3}{12(1+\nu)},\quad \mu=\frac{\nu Ea^3}{12(1-\nu^2)},$$

where E is the Young modulus, ν the Poisson ratio of the elastic material of the plate and a the thickness of the plate. □

Problems

5.2 Prove (5.8).

5.3 Show that the analogue of Stokes equations (5.6) in a two-dimensional simply connected domain Ω can be formulated as the biharmonic problem (2.22) by introducing the stream function as unknown.

5.4 Show that the plate problem (5.9)–(5.11) takes the form (2.22) if $\Gamma_1=\Gamma$.

6. Direct methods for solving linear systems of equations

6.1 Introduction

We have seen in Chapters 1 and 2 that application of the finite element method to a linear elliptic problem typically leads to a linear system of equations

$$(6.1) \qquad A\xi = b,$$

where $A = (a_{ij})$ is a symmetric, positive definite and sparse $M \times M$ matrix, and $b \in R^M$. We also know that the unique solution $\xi \in R^M$ of (6.1) can be equivalently characterized as the solution of the quadratic minimization problem

$$(6.2) \qquad \min_{\eta \in R^M} \left[\frac{1}{2}\,\eta \cdot A\eta - b \cdot \eta\right].$$

To compute the solution ξ we can start either from (6.1) or (6.2). In this chapter we shall study some *direct* methods, or methods based on Gaussian elimination, for the solution of (6.1). In the next chapter we shall study some *minimization algorithms* for the solution of (6.2) that may be viewed equivalently as iterative methods for (6.1).

Remark Finite element methods for first order hyperbolic problems typically lead to non-symmetric linear systems of equations, see Chapter 9 below. In this case there is no associated minimization problems (unless a least-squares formulation is used) and it is not yet clear how to construct efficient iterative methods for general classes of non-symmetric problems. Thus, for such problems Gaussian elimination (with pivoting, cf below) is often used. □

6.2 Gaussian elimination. Cholesky's method

We recall (cf any basic course in numerical analysis) that using Gaussian elimination to solve (6.1), we obtain a LU-*factorization* of A of the form

(6.3) $\qquad A=LU,$

where $L=(l_{ij})$ is a *lower triangular* $M\times M$ matrix (i e, $l_{ij}=0$ if $j>i$), and $U=(u_{ij})$ is an *upper triangular* matrix (i e, $u_{ij}=0$ if $j<i$), or diagrammatically:

$$L=\begin{bmatrix} x & & & & \\ x & x & & 0 & \\ x & x & x & & \\ \vdots & & \ddots & x & \\ x & \cdots & \cdots & x & x \end{bmatrix}, \quad U=\begin{bmatrix} x & x & x & \cdots & x \\ & x & x & \cdots & x \\ & & x & \ddots & \vdots \\ & 0 & & x & x \\ & & & & x \end{bmatrix}.$$

From the factorization (6.3) it is easy to solve the system $A\xi=b$ by using *forward* and *backward substitution* to solve the triangular systems:

(6.4a) $\qquad L\eta=b,$

(6.4b) $\qquad U\xi=\eta.$

We recall that $U=A^{(M)}$ where the matrices $A^{(k)}$, $k=1,\ldots,M$, are successively computed as follows:

(i) $\qquad A^{(1)}=A,$

(ii) $\qquad$ Given $A^{(k)}$ of the form

$$A^{(k)}=\begin{bmatrix} a_{11}^{(k)} & & & & a_{ln}^{(k)} \\ 0 & & & & \\ & & 0 & a_{kk}^{(k)} & a_{kn}^{(k)} \\ & & & & \\ 0 & & 0 & a_{nk}^{(k)} & a_{nn}^{(k)} \end{bmatrix}$$

determine $A^{(k+1)}=(a_{ij}^{(k+1)})$ as follows

(6.5)
$$a_{ij}^{(k+1)}=a_{ij}^{(k)}, \qquad i=1,\ldots,k, \text{ or } j=1,\ldots,k-1,$$
$$a_{ij}^{(k+1)}=a_{ij}^{(k)}-\frac{a_{ik}^{(k)}}{a_{kk}^{(k)}}a_{kj}^{(k)} \qquad i=k+1,\ldots,M, \text{ and } j=k,\ldots,M,$$

under the assumption that $a_{kk}^{(k)}\neq 0$.

We also recall that $L=(l_{ij})$, where

$$\begin{cases} l_{ii}=1, & i=1,\ldots,M, \\ l_{ik}=-\dfrac{a_{ik}^{(k)}}{a_{kk}^{(k)}}, & i=k+1,\ldots,M, \qquad k=1,\ldots,M, \\ l_{ik}=0, & \text{if } i<k. \end{cases}$$

One can show that if A is symmetric positive definite, then $a_{kk}^{(k)}>0$, $k=1, \ldots, M$. Thus, Gaussian elimination can be performed without pivoting. In addition, under the same hypothesis it is not necessary to perform pivoting to prevent numerical instability due to too small pivot elements $a_{kk}^{(k)}$. Thus, we may perform the Gaussian elimination in any desired order. We will see below that different direct methods for (6.1) essentially differ in the choice of the order of the elimination, i e, the enumeration of the nodes in case we perform the elimination according to the ordering of the nodes.

Since A is symmetric positive definite we may alternatively factor A as

$$A=BB^T,$$

with $B=DL$ and where D is a diagonal matrix with diagonal elements

$$d_{kk}=\sqrt{a_{kk}^{(k)}}, \quad k=1, \ldots, M,$$

and L and $a_{kk}^{(k)}$ are obtained through the Gaussian elimination given above. Here B^T denotes the transpose of the matrix B. The elements b_{ij} of the matrix B can alternatively be determined using *Cholesky's method* as follows:

$$b_{11}=\sqrt{a_{11}},$$

$$b_{i1}=\frac{a_{i1}}{b_{11}}, \qquad i=2, \ldots, M,$$

and for $j=2, \ldots, M$,

$$b_{jj}=\left[a_{jj}-\sum_{k=1}^{j-1} b_{jk}^2 \right]$$

$$b_{ij}=(a_{ij}-\sum_{k=1}^{j-1} b_{ik}b_{jk})/b_{jj}, \quad i=j+1, \ldots, M.$$

6.3 Operation counts. Band matrices

The number of arithmetic operations to obtain an LU-factorization of a *dense* $M\times M$ matrix (i e matrix with few zero elements) is asymptotically of the order $M^3/3$. If the matrix is sparse, then it is possible to greatly reduce the number of operations by using the sparsity. This is particularly easy to do if the matrix A is a *band matrix,* i e, there is a natural number d, the *band width,* such that

$$a_{ij}=0 \text{ if } |i-j|>d.$$

A band matrix has the following form, where the shaded area indicates where non-zero elements may occur (some elements in the band may be zero):

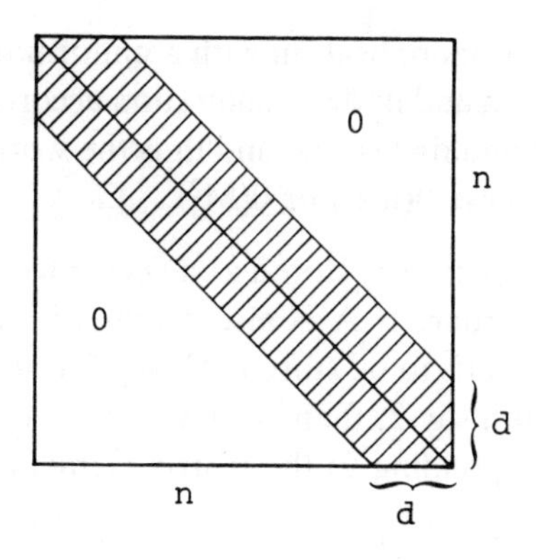

To factor an M×M matrix with band width d one needs asymptotically $Md^2/2$ operations (cf Problem 6.1), which is much less than the number $M^3/3$ for a dense matrix if d is much smaller than M.

In our applications when $a_{ij}=a(\varphi_i, \varphi_j)$, where $a(. , .)$ is a bilinear form and $\{\varphi_1, . . ., \varphi_M\}$ is a basis for a finite element space V_h, we have that

$$d=\max \{|i-j|: \varphi_i \text{ and } \varphi_j \text{ are associated with degrees of freedom belonging to the same element}\}.$$

Clearly, the band width depends on the chosen enumeration of the nodes, and thus if Gaussian elimination is to be used, then we want to enumerate the nodes so as to make the band width (nearly) as small as possible.

Example 6.1 Let us consider the following enumeration giving minimal band width

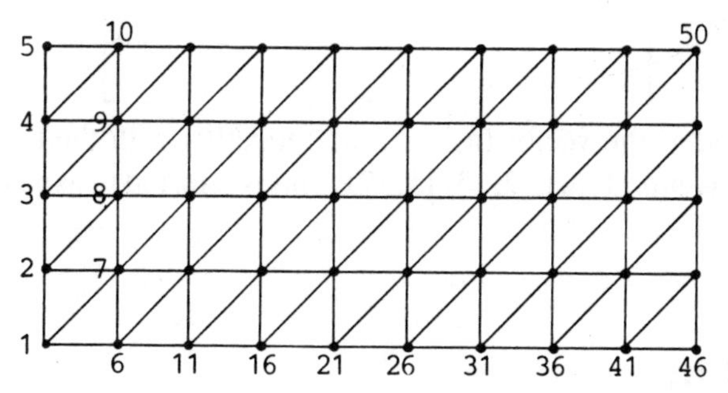

In this case we have d=6 (assuming one degree of freedom associated with each node). With a horizontal enumeration instead we would have d=11. □

Example 6.2 In a typical application with a uniform triangulation of the unit square with mesh size h and node enumeration according to Example 6.1, we have that $M=0(h^{-2})$ and $d=0(h^{-1})$, and thus the work estimate for Gaussian elimination is in this case $0(h^{-4})$ or $0(M^2)$. □

Note that a band matrix A is stored as a vector with e g the columns in the band in consecutive order. If A is also symmetric then only e g the upper triangular part of A needs to be stored. Thus, if $A=(a_{ij})$ is a symmetric band matrix with band width say 2, then A may be stored as a vector $a=(a_i)$ with the elements a_i corresponding to the matrix elements a_{ij} as follows:

$$A=\begin{bmatrix} a_1 & a_2 & a_4 & 0 & 0 & 0 \\ & a_3 & a_5 & a_7 & 0 & 0 \\ & & a_6 & a_8 & a_{10} & 0 \\ \text{sym} & & & a_9 & a_{11} & a_{13} \\ & & & & a_{12} & \cdot \\ & & & & & \cdot \end{bmatrix}.$$

Remark It is sometimes convenient to allow the band width to vary from one column to another. To store A in this case, we again store the columns of the band consecutively in a vector $a=(a_i)$. We then also have to supply information concerning the indices of the diagonal elements. As an example, a matrix A with the following variable band structure

$$A=\begin{bmatrix} a_1 & a_2 & a_4 & 0 & 0 & 0 & 0 \\ & a_3 & a_5 & a_7 & a_{10} & 0 & 0 \\ & & a_6 & a_8 & a_{11} & 0 & 0 \\ & & & a_9 & a_{12} & 0 & 0 \\ \text{sym} & & & & a_{13} & a_{14} & 0 \\ & & & & & a_{15} & a_{16} \\ & & & & & & a_{17} \end{bmatrix}$$

can be stored as the vector $(a_1, \ldots, a_{17})$ together with the list of indices of diagonal elements (1, 3, 6, 9, 13, 15, 17). This is referred to as a *skyline* method of storage. □

6.4 Fill-in

Using (6.5) it is easy to see that if A is a band matrix with band width d, then so are the factors L and U in an LU-factorization of A. However, the matrices L and U may have non-zero elements within the band at locations where the elements of A are zero. This is called *fill-in.* In the applications most of the elements of A within the band are zero (see e g (1.25)), while with usual orderings such as in Example 6.1, most of the elements of the factors L and U within the band are non-zero. Thus, the factors L and U contain many more non-zero elements than A and we have a considerable fill-in. Different enumerations of the nodes may give different degrees of fill-in, cf the nested dissection method below. Notice that the density of the factors L and U influence the cost of the backward and forward substitutions (6.4a, b). With most of the elements non-zero within the band, as is typical with usual orderings, this cost is $0(Md)$.

We will now briefly consider some common variants of Gaussian elimination, namely the *frontal method* (cf [I]) and *nested dissection* (cf [Ge]).

6.5 The frontal method

In this method the assembly of the stiffness matrix and the Gaussian elimination are carried out in parallel. Moreover, it is not necessary to store the entire matrices $A^{(k)}$ obtained through the elimination process in the fast memory, which may be difficult if M is large; instead it is sufficient at each step of the elimination to store just a smaller part of $A^{(k)}$ in the fast memory and communicate with a secondary memory only at the beginning and end of each step.

Let us give some more details of this procedure and to be specific let us consider the same situation as in Section 1.8. That is, let $A=(a_{ij})$ be the stiffness matrix associated with the Neumann problem of Example 2.1 and the standard finite element space of piecewise linear functions on a triangulation $T_h=\{K\}$ with basis $\{\varphi_1, \ldots, \varphi_M\}$. Suppose further that the nodes are enumerated so that A is a band matrix with band width $d<M$. The frontal method is based on the following facts:

(i) The matrices $A^{(k)}$, $k=1, \ldots, M$, obtained through the Gaussian elimination, are all band matrices with band width d. To compute $A^{(k+1)}$ with $A^{(k)}$ given, we need to change the elements $a_{kj}^{(k)}$ by subtracting the quantities

$$\frac{a_{ik}^{(k)}}{a_{kk}^{(k)}} a_{kj}^{(k)} \quad \text{for } i=k+1, \ldots, k+d, \; j=k, k+1, \ldots, k+d.$$

In other words, to compute $A^{(k+1)}$ it is sufficient to work with the $(d+1)\times(d+1)$ matrix

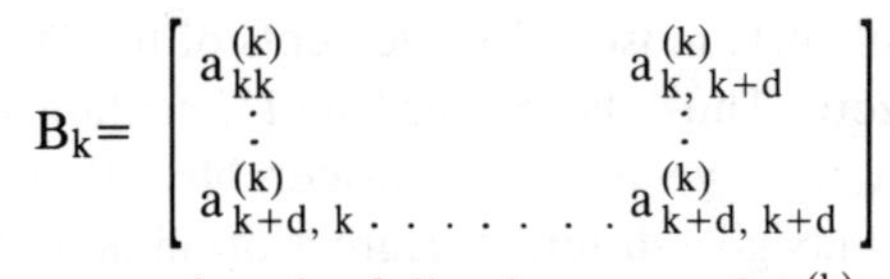

$$B_k=\begin{bmatrix} a^{(k)}_{kk} & & a^{(k)}_{k,\,k+d} \\ \vdots & & \vdots \\ a^{(k)}_{k+d,\,k} & \cdots\cdots & a^{(k)}_{k+d,\,k+d} \end{bmatrix}$$

occupying the following part of $A^{(k)}$:

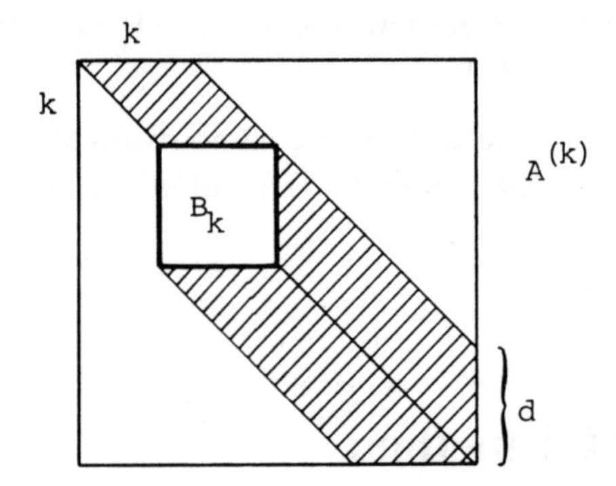

(ii) In the assembly

$$a_{ij}=\sum_K a_K(\varphi_i,\ \varphi_j),$$

we add the contributions $a_K(\varphi_i,\ \varphi_j)$ from triangles K in which both node i and node j are vertices. Now, to eliminate the variable ξ_k, i e, to take the step from $A^{(k)}$ to $A^{(k+1)}$, we only need to have the matrix elements in column and row k fully assembled, while the matrix element a_{ij} with $i,\ j\geqslant k+1$ may be modified at a later stage by adding the contributions $a_K(\varphi_i,\varphi_j)$ not yet included.

From (i) and (ii) it follows that we may perform the assembly and elimination in parallel. In step k with $A^{(k)}$ given, we first assemble all remaining contributions from triangles K with node k as vertex, and then we compute $A^{(k+1)}$ in the usual way. In this case only the elements in B_k, the so-called *active area,* will be modified. At the end of step k we store row k of $A^{(k)}$ (or $A^{(k+1)}$), which will be row k of the upper triangular factor U in the factorization A=LU, in a secondary memory and then move the active area one step in the south-east direction.

The line dividing the triangles with fully assembled contributions and the remaining triangles with not yet fully assembled contributions, is called the *front*. The assembly activity takes place at the front and with a suitable enumeration the front will sweep over the region Ω in the combined assembly-elimination. We now consider an example.

Example 6.3 Consider the following triangulation of the region Ω:

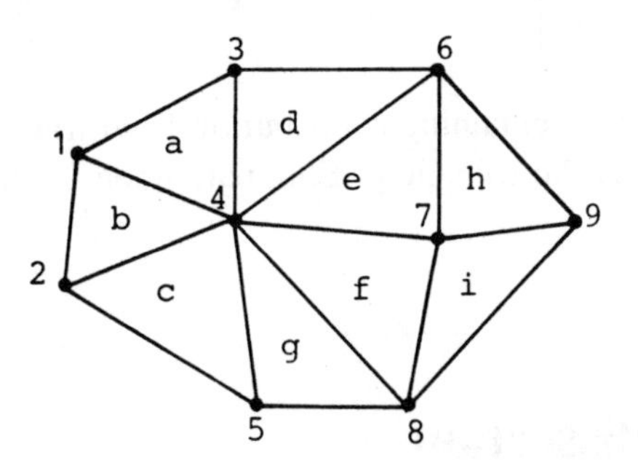

where the nodes have been numbered and the triangles are denoted by the letters a–i. The corresponding stiffness matrix has the following structure where x indicates non-zero elements.

active area at step 1

$$
\left[\begin{array}{cccc|ccccc}
x & x & x & x & & & & & \\
x & x & & x & x & & & & \\
x & & x & x & & x & & & \\
x & x & x & x & x & x & x & x & \\
\cline{1-4}
 & x & & x & x & & & x & \\
 & & x & x & & x & x & & x \\
 & & & x & & x & x & x & x \\
 & & & x & x & & x & x & x \\
 & & & & & x & x & x & x
\end{array}\right]
$$

Step 1. Assemble contributions from triangles with node 1 as vertex, i e, the triangles a and b. Eliminate node 1 (variable ξ_1) and store row 1.

Let x_1 denote the elements modified in Step 1. We have now obtained the following situation (note the fill-in: the element at location 23 is now non-zero corresponding to the fact that node 2 now is coupled to node 3 through the eliminated node 1), active area at step 2

$$
\begin{bmatrix}
x & x & x & x & & & & & \\
x_1 & x_1 & x_1 & x_1 & & & & & \\
x_1 & x_1 & x_1 & x_1 & & x & & & \\
x_1 & x_1 & x_1 & x_1 & x & x & x & x & \\
 & x & & x & x & & & x & \\
 & & x & x & & x & x & & x \\
 & & & x & & x & x & x & x \\
 & & & x & x & & x & x & x \\
 & & & & & x & x & x & x
\end{bmatrix}
$$

Step 2. Assemble the remaining contribution from triangles with node 2 as a vertex, i e, the triangle c. Eliminate node 2, etc.

6.6 Nested dissection

In the nested dissection method one uses an enumeration of the nodes radically different from the ones we have used above. We illustrate the method in a simple example with the finite element method of the previous subsection on the following triangulation of the unit square Ω:

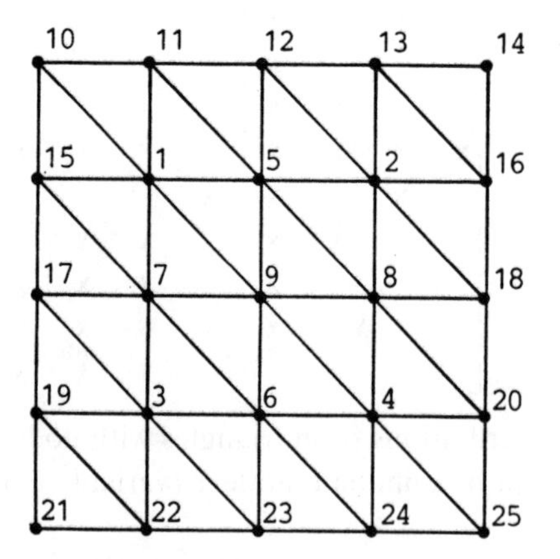

Step 1. We first view the structure or triangulation of Ω subdivided into four *substructures* A-D as follows:

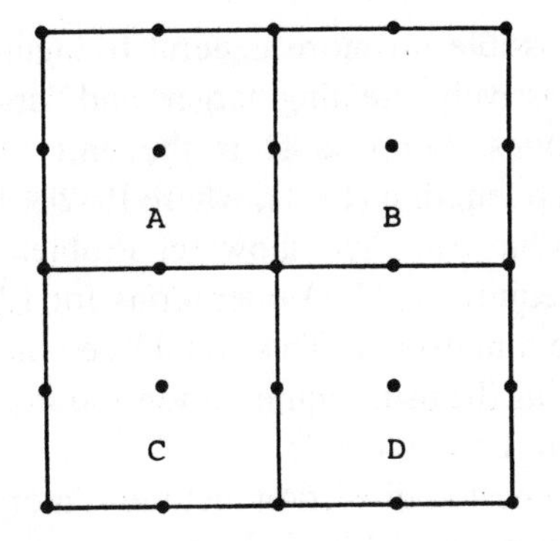

We then eliminate the *inner* nodes in each substructure, i e the nodes 1 to 4.

Step 2. We now combine A and B into one structure AB, and C and D into one structure DC:

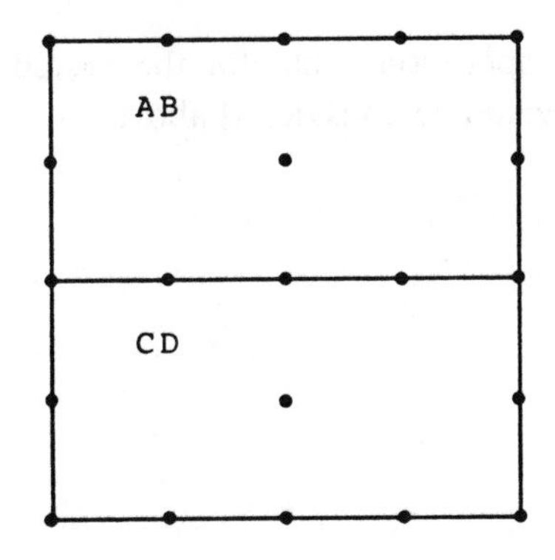

We then eliminate the inner nodes in AB and CD, and combine AB with CD into one structure ABCD:

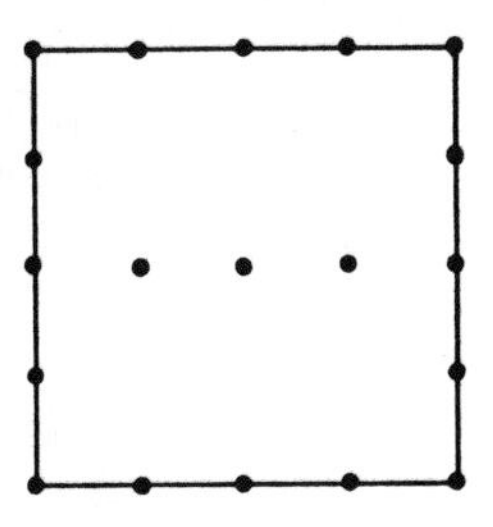

and eliminate the inner nodes 7–9.

Step 4. The nodes 10–25 are eliminated.

Analogously, it is possible on more general triangulations to perform the elimination by successively creating larger and larger substructures and eliminating inner nodes. Suppose Ω is the unit square with a uniform triangulation with step length $h/(P-1)$, where $P=2^p+1$, p a natural number, with $M=P^2$ nodes. One can then show (cf Problem 6.2) that the nested dessection method requires $0(M^{3/2})$ operations for LU-factorization of the corresponding stiffness matrix A. This should be compared with the $0(M^2)$ operations needed using the usual enumeration and storing A as a band matrix with band width M (cf Example 6.2).

The reason that the nested dissection method is more efficient in this case, is the fact that it produces less fill-in. For general geometries, however, it may be rather difficult to implement the nested dissection method.

Problems

6.1 Show that the number of operations to factor a $M\times M$ matrix with band width d, is of the order $Md^2/2$.

6.2 Show that the operation count for the nested dissection method is $0(M^{3/2})$ in the example considered above.

7. Minimization algorithms. Iterative methods

7.1 Introduction

In this chapter we consider *iterative methods* for the numerical solution of minimization problems of the form

$$\text{(7.1)} \qquad \operatorname*{Min}_{\eta\in R^M} f(\eta),$$

where $f: R^M \to R$ is a quadratic function

$$\text{(7.2)} \qquad f(\eta)=\frac{1}{2}\,\eta\cdot A\eta-b\cdot\eta,$$

with A a sparse symmetric positive definite $M\times M$ matrix and $b\in R^M$. As we have seen above, application of the finite element method to a linear elliptic problem typically leads to a problem of the form (7.1). We know that (7.1) admits a unique solution $\xi\in R^M$ equivalently characterized by the equation

$$\text{(7.3)} \qquad A\xi=b.$$

Iterative methods for the solution of (7.1), or equivalently (7.3), play an increasingly important role in finite element applications. A key fact making iterative methods advantageous is the extreme sparsity of the matrix A in standard applications. For a given type of finite element the number of non-zero entries in each row of A is bounded independently of the mesh size. This means that if only the non-zero entries of A are stored, then to compute $A\eta$ for a given $\eta\in R^M$ takes $0(M)$ operations (compared to $0(M^2)$ if A is full). We emphasize that to achieve the operation count $0(M)$ we may not store A as e g a band matrix; only the non-zero entries of A should be stored, (cf Remark 7.3 below).

We will consider iterative methods or *minimization algorithms* for (7.1) of the form: Given an initial approximation $\xi^0\in R^M$ of the exact solution ξ, find successive approximations $\xi^k\in R^M$, $k=1, 2, \ldots$, of the form

$$\text{(7.4)} \qquad \xi^{k+1}=\xi^k+\alpha_k d^k, \quad k=0, 1, \ldots,$$

where $d^k \in R^M$ is a *search direction* and $\alpha_k > 0$ is a *step length* (note that the summation convention is not used in this chapter). Different methods differ in the choice of the search direction d^k and step length α_k. We will consider (a) the *gradient method,* and (b) *the conjugate gradient method* together with so-called *preconditioned* variants of these methods.

We use the following notation. Given a smooth function $g: R^M \rightarrow R$, denote by g' or ∇g the *gradient* of $g = g(\eta)$, i e,

$$g' = \nabla g = \left(\frac{\partial g}{\partial \eta_1}, \frac{\partial g}{\partial \eta_2}, \ldots, \frac{\partial g}{\partial \eta_M} \right).$$

Further, define the *Hessian* of g to be the M×M matrix $g'' = (g_{,ij})$, i e,

$$g'' = \begin{bmatrix} \frac{\partial^2 g}{\partial \eta_1^2} & \cdots\cdots & \frac{\partial^2 g}{\partial \eta_1 \partial \eta_M} \\ & & \\ \frac{\partial^2 g}{\partial \eta_M \partial \eta_1} & \cdots\cdots & \frac{\partial^2 g}{\partial \eta_M^2} \end{bmatrix}.$$

For the quadratic function f of (7.2), we have

$$f'(\eta) = A\eta - b, \qquad \eta \in R^M,$$

and

$$f''(\eta) = A, \qquad \eta \in R^M.$$

With ξ^{k+1} given by (7.4), we have by Taylor's formula

$$g(\xi^{k+1}) = g(\xi^k) + \alpha_k g'(\xi^k) \cdot d^k + \frac{\alpha_k^2}{2} d^k \cdot g''(\eta) d^k,$$

where η lies on the line segment between ξ^k and ξ^{k+1}. If the elements in g'' are bounded in a neighborhood of ξ^k, we thus have

$$g(\xi^{k+1}) = g(\xi^k) + \alpha_k g'(\xi^k) \cdot d^k + 0(\alpha_k^2), \text{ as } \alpha \rightarrow 0.$$

It follows that if

$$g'(\xi^k) \cdot d^k < 0, \tag{7.5}$$

then $g(\xi^{k+1}) < g(\xi^k)$ if α_k is sufficiently small. With this motivation we say that d^k is a *descent direction* for g if (7.5) holds, because then g will decrease if we move a small distance from ξ^k in the direction d^k. In particular, (7.5) holds if we choose (see Fig 7.1)

$$d^k = -g'(\xi^k) \tag{7.6}$$

and if $g'(\xi^k) \neq 0$. In this case (7.4) corresponds to one step of the *gradient method* or the *steepest descent* method for the minimization problem $\min_{\eta \in R^M} g(\eta)$. To choose the step-length α_k we may, for example, determine α_k so that

$$g(\xi^k + \alpha_k d^k) = \min_{\alpha \geqslant 0} g(\xi^k + \alpha d^k),$$

in which case α_k is said to be *optimal*. To determine α_k we perform a one-dimensional *line-search* to minimize g in the direction d^k starting from ξ^k. If α_k is optimal, then $\frac{d}{d\alpha} g(\xi^k + \alpha d^k) = 0$ for $\alpha = \alpha_k$ so that (see Fig 7.2).

$$g'(\xi^{k+1}) \cdot d^k = 0. \tag{7.7}$$

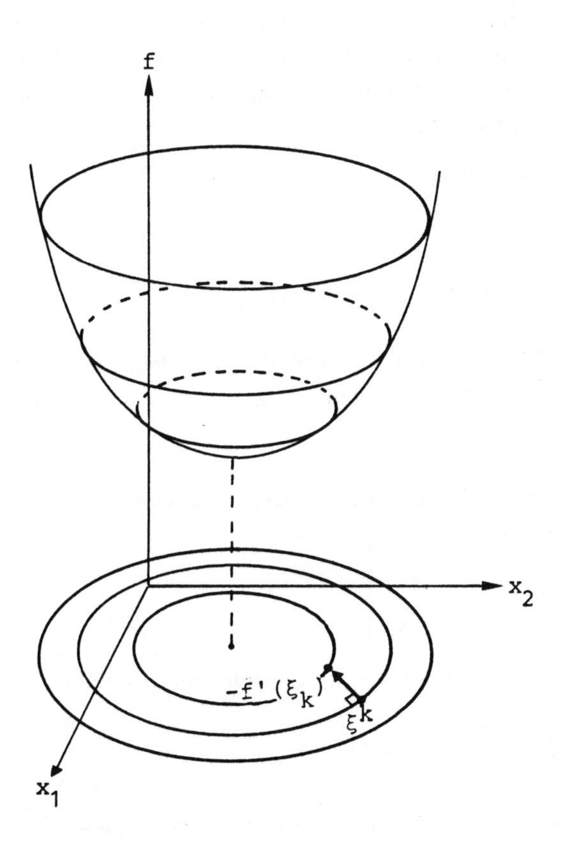

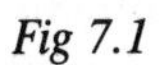
Fig 7.1

In particular, if g is the function f given by (7.2), then by (7.7)

$$0=f'(\xi^k+\alpha_k d^k)\cdot d^k=(A(\xi^k+\alpha_k d^k)-b)\cdot d^k$$
$$=(A\xi^k-b)\cdot d^k+\alpha_k d^k\cdot Ad^k,$$

so that in this case α_k is given by the following simple formula:

$$\alpha_k=-\frac{(A\xi^k-b)\cdot d^k}{d^k\cdot Ad^k}. \tag{7.8}$$

Remark 7.1 Note that $g'(\xi^k)$ is orthogonal to a level curve for g through ξ^k (a level curve for g is a curve γ: $[a, b]\rightarrow R^M$ such that $g(\gamma(t))=\text{constant}$ for $t\in[a, b]$, see Fig 7.1). □

We will be particularly interested in the *rate of convergence* of the different methods to be studied, i e, we will be interested in estimating how many steps or iterations of the form (7.4) will be needed to reduce the initial error $\xi-\xi^0$ by a certain factor. We will then see that the rate of convergence depends on the *condition number* $\varkappa(A)$ of A defined by:

$$\varkappa(A)=\frac{\lambda_{max}}{\lambda_{min}}, \tag{7.9}$$

where

$$\lambda_{max}=\max_j \lambda_j, \quad \lambda_{min}=\min_j \lambda_j,$$

and λ_j, $j=1, \ldots, M$, are the (positive) eigenvalues of A. We assume that the eigenvalues are ordered so that $\lambda_1\leq\lambda_2\leq \ldots \lambda_M$, in which case of course $\lambda_{min}=\lambda_1$ and $\lambda_{max}=\lambda_M$.

Example 7.1 Consider the special case of (7.1) with A the 2×2 diagonal matrix

$$A=\begin{bmatrix} \lambda_1 & 0 \\ 0 & \lambda_2 \end{bmatrix},$$

where $0<\lambda_1<\lambda_2$ and $b=0$, i e, we consider the problem

$$\underset{\eta\in R^2}{\text{Min}}\frac{1}{2}(\lambda_1\eta_1^2+\lambda_2\eta_2^2), \tag{7.10}$$

with solution $\xi=0$. The level curves of f are in this case ellipses with half-axis proportional to $\sqrt{1/\lambda_1}$ and $\sqrt{1/\lambda_2}$. The sequence $\xi^0, \xi^1, \ldots,$ obtained by applying the gradient method with optimal step length to (7.10) is plotted in

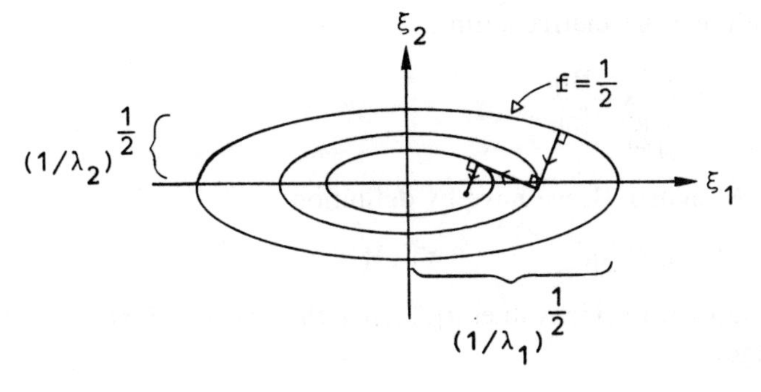

Fig 7.2

Fig 7.2. We see that as the condition number becomes larger and the level curves more elongated, the sequence $\xi^0, \xi^1, \ldots,$ has a more pronounced zig-zag and convergence becomes slower. □

The above example shows that it is important to understand the behaviour of the condition number $\varkappa(A)$. We will see that in a typical case when A results from application of the finite element method to a second order elliptic problem (such as e g the Poisson equation (1.16)), then

(7.11) $\qquad \varkappa(A)=0(h^{-2}),$

where as usual h is the mesh parameter. For a problem of order four such as e g the biharmonic problem (2.22), one has $\varkappa(A)=0(h^{-4})$. To be more precise, these estimates hold if the finite element mesh is quasi-uniform, i e, all elements have roughly the same size (cf (7.44a) below), and if the usual minimum angle assumption (4.1) is valid.

We finally conclude our preparations by recalling that

(7.12a) $$\lambda_{min} = \min_{\substack{\eta \in R^M \\ \eta \neq 0}} \frac{\eta \cdot A\eta}{|\eta|^2},$$

(7.12b) $$\lambda_{max} = \max_{\substack{\eta \in R^M \\ \eta \neq 0}} \frac{\eta \cdot A\eta}{|\eta|^2},$$

where $|\cdot|$ denotes the usual Euclidean norm

$$|\eta| = \left(\sum_{i=1}^{M} \eta_i^2\right)^{1/2}.$$

Further, defining the matrix norm

$$|B| = \max_{\substack{\eta \in R^M \\ \eta \neq 0}} \frac{|B\eta|}{|\eta|},$$

for the M×M matrix B, we have by definition

$$(7.13) \qquad |B\eta| \leq |B|\,|\eta| \qquad \forall \eta \in R^M.$$

If B is symmetric with eigenvalues $\mu_1, \ldots, \mu_M$, then we have (cf any basic course in linear algebra):

$$(7.14) \qquad |B| = \max_j |\mu_j|.$$

7.2 The gradient method

We will now study the rate of convergence of the gradient method for (7.1) with *constant step length,* i e, the method

$$(7.15) \qquad \xi^{k+1} = \xi^k + \alpha d^k, \qquad k=0, 1, \ldots,$$

$$d^k = -f'(\xi^k) = -(A\xi^k - b).$$

Here α is a suitably chosen (sufficiently small) positive constant. The appropriate size of α will become clear through the following analysis. Since the exact solution satisfies $A\xi = b$, we have

$$(7.16) \qquad \xi = \xi - \alpha(A\xi - b),$$

which after subtraction with (7.15) gives the following relation for the error $e^k = \xi - \xi^k$:

$$(7.17) \qquad e^{k+1} = (I - \alpha A)e^k, \qquad k=0, 1, \ldots,$$

where I is the identity matrix. Thus, by (7.13) we have

Thus, by (7.13) we have

$$(7.18) \qquad |e^{k+1}| \leq |I - \alpha A|\,|e^k|.$$

We would now like to be able to guarantee that

$$(7.19) \qquad |I - \alpha A| \equiv \gamma < 1.$$

In this case the error would get reduced by the factor γ at each step and the smaller γ is, the more rapid is the convergence. Now, by (7.14) we have

$$|I - \alpha A| = \max_j |1 - \alpha\lambda_j|,$$

so that (7.19) holds if and only if

$$1-\alpha\lambda_j>-1, \; j=1, \ldots, M,$$

since the λ_j are positive. We thus conclude that α has to be chosen so that $\alpha\lambda_{max}<2$. Choosing now $\alpha=1/\lambda_{max}$, which is close to the best choice, we have

$$|I-\alpha A|=1-\frac{\lambda_{min}}{\lambda_{max}}=1-\frac{1}{\varkappa(A)}.$$

From (7.18) we thus have

$$|e^{k+1}|\leqslant\gamma|e^k|$$

with $\gamma=1-\frac{1}{\varkappa(A)}$, and by induction

$$|e^k|\leqslant\gamma^k|e^0|, \qquad k=1, \ldots, .$$

Let us now estimate the number of steps n required to reduce the initial error $|e^0|$ by a certain given factor $\varepsilon>0$. That is, we seek the smallest n such that

$$(7.20) \qquad \left(1-\frac{1}{\varkappa(A)}\right)^n\leqslant\varepsilon,$$

or equivalenty

$$-n \log \left(1-\frac{1}{\varkappa(A)}\right)\geqslant\log\frac{1}{\varepsilon}.$$

Using the easily proved fact that $-\log(1-x)>x$ for $x<1$, we see that (7.20) is satisfied if

$$(7.21) \qquad n\geqslant\varkappa(A)\log\frac{1}{\varepsilon}.$$

We conclude that the required number of iterations in the gradient method (7.15), with a suitably chosen constant step α, is proportional to the condition number $\varkappa(A)$ and the number of decimals in the error reduction factor ε. In a typical FEM application involving a second order elliptic equation we have that $\varkappa(A)=0(h^{-2})$ and in this case we would have $n=0(h^{-2})$, ie, a very large number of iterations would have to be performed.

By using an eigenvector expansion, it is possible to see more clearly why the gradient method is not efficient if $\varkappa(A)$ is large. To this end, let $\psi_1, \ldots, \psi_M\in R^M$ be the orthonormal basis of eigenvectors corresponding to the eigenvalues $\lambda_1, \ldots, \lambda_M$ of A, ie,

$$A\psi_j=\lambda_j\psi_j, \qquad j=1, \ldots, M.$$

Expanding the error e^k in the basis $\psi_1, \ldots, \psi_M$, we have

$$e^k = \sum_{j=1}^{M} e_j^k \psi_j, \qquad e_j^k = e^k \cdot \psi_j,$$

and the relation (7.17) takes the form

$$e_j^{k+1} = \left(1 - \frac{\lambda_j}{\lambda_{max}}\right) e_j^k \qquad j = 1, \ldots, M, \tag{7.22}$$

with $\alpha = 1/\lambda_{max}$. The relation (7.22) gives the error reduction for each component e_j^k of the error e^k. As the λ_j are ordered in increasing order with $\lambda_{max} = \lambda_M$, we see from (7.22) that for j such that $\lambda \geq \lambda_{max}/2$ (ie for "big" j), the corresponding component e_j^k gets reduced by at least a factor $1/2$ at each step and a considerable reduction takes place. On the other hand, for "small" j, λ_j/λ_{max} is much smaller than 1, the error reduction factor $(1 - \lambda_j/\lambda_{max})$ is close to one and the reduction is small. Thus, error components e_j^k for large j are reduced quickly, while components e_j^k for j small are only slowly reduced. Another way of saying this is that highly oscillatory components of the error are quickly reduced while more slowly varying components only get slowly reduced. This is because the eigenvectors ψ_j for large j are rapidly oscillating and for small j the ψ_j vary "more smoothly", cf Problem 8.1.

To sum up, we may say that the gradient method efficiently reduces highly oscillatory components of the error while the smooth components only become small very slowly, and thus as a whole the gradient method is inefficient (cf Section 7.5 below on multi-grid methods where the gradient method is put to very efficient use.)

Remark 7.2 The gradient method for (7.1) with the optimal step length of (7.8) is given by:

$$\xi^{k+1} = \xi^k - \alpha_k (A\xi^k - b), \tag{7.23a}$$

$$\alpha_k = \frac{r^k \cdot r^k}{r^k \cdot Ar^k}, \quad r^k = A\xi^k - b. \tag{7.23b}$$

The convergence properties of this method are similar to those of the gradient method with constant steps just studied and in particular the required number of iterations is proportional to $\varkappa(A)$, cf Problem 7.2. □

7.3 The conjugate gradient method

We will now describe a more efficient iterative method for (7.1) of the form (7.4), namely the *conjugate gradient method.* In this method the step length α_k is chosen to be optimal and the search directions d^k are *conjugate,* i e,

(7.24) $\qquad d^i \cdot Ad^j=0, \quad i\neq j.$

Since A is positive definite, we may define a scalar product $<.\,,.>$ on R^M by

$$<\zeta, \eta>=\zeta \cdot A\eta, \qquad \zeta, \eta \in R^M,$$

and (7.24) can then be written

$$<d^i, d^j>=0, \qquad i\neq j.$$

The norm $\|\cdot\|_A$ corresponding to $<.\,,.>$ is the *energy* norm:

$$\|\eta\|_A=<\eta, \eta>^{1/2}, \qquad \eta\in R^M.$$

The conjugate gradient method can now be stated as follows: Given $\xi^0\in R^M$ and $d^0=-r^0$, find ξ^k and d^k, $k=1, 2, \ldots$, such that

(7.25a) $\qquad \xi^{k+1}=\xi^k+\alpha_k d^k,$

(7.25b) $\qquad \alpha_k=-\dfrac{r^k \cdot d^k}{<d^k, d^k>},$

(7.25c) $\qquad d^{k+1}=-r^{k+1}+\beta_k d^k,$

(7.25d) $\qquad \beta_k=\dfrac{<r^{k+1}, d^k>}{<d^k, d^k>},$

where

$$r^k=f'(\xi^k)=A\xi^k-b.$$

If we compare with (7.8) we see that (7.25b) means that α_k is optimal. We note that the new search direction d^{k+1} is a linear combination of the new gradient r^{k+1} and the old search direction d^k. Further, in view of (7.25c), the condition $<d^{k+1}, d^k>=0$ is equivalent to

$$<-r^{k+1}+\beta_k d^k, d^k>=0,$$

which is the same as (7.25d). We thus see directly that the new search direction d^{k+1} is conjugate with respect to the old direction d^k. We now take an essential step in the analysis of the method and prove that d^{k+1} is also conjugate with respect to all other previous search directions d^j, $j=0, \ldots, k$ (cf. [Lu]). We will need the following lemma, where we use the notation

$$[\eta^0, \ldots, \eta^m]=\{\eta \in R^M: \eta=\sum_{j=0}^{m} a_j\eta^j,\ a_j \in R\}$$

$$=\text{linear space spanned by } \eta^j \in R^M,\ j=0, \ldots, m.$$

Lemma 7.1 For $m=0, 1, \ldots$, we have $[d^0, \ldots, d^m]=[r^0, \ldots, r^m]$ $=[r^0, Ar^0, \ldots, A^m r^0]$.

Proof We use an induction argument. The stated equality clearly holds for $m=0$. Suppose now that the equality holds for $m=k$. We first observe that after multiplication by A, (7.25a) gives

$$r^{k+1}=r^k+\alpha_k Ad^k. \tag{7.26}$$

By the induction assumption, we have $d^k \in [r^0, Ar^0, \ldots, A^k r^0]$ so that $Ad^k \in$ $[r^0, Ar^0, \ldots, A^{k+1}r^0]$ which shows that

$$[r^0, \ldots, r^{k+1}] \subset [r^0, Ar^0, \ldots, A^{k+1}r^0]. \tag{7.27}$$

On the other hand, according to the induction hypothesis, we have $A^k r^0 \in [d^0, \ldots, d^k]$ so that $A^{k+1}r^0 \in [Ad^0, \ldots, Ad^k]$ which together with (7.26) shows that $A^{k+1}r^0 \in [r^0, \ldots, r^{k+1}]$. Thus we have $[r^0, Ar^0, \ldots, A^{k+1}r^0] \subset [r^0, \ldots, r^{k+1}]$ which by (7.27) shows that $[r^0, Ar^0, \ldots, A^{k+1}r^0]=[r^0, \ldots, r^{k+1}]$. Finally, from (7.25c) we clearly have that $[r^0, \ldots, r^{k+1}]=[d^0, \ldots, d^{k+1}]$ and the induction step is thereby complete. □

We can now prove

Lemma 7.2 The search directions d^i are pairwise conjugate, i e,

$$\langle d^i, d^j \rangle=0, \qquad i \neq j. \tag{7.28}$$

Further, the gradients r^i are orthogonal, i e,

$$r^i \cdot r^j=0, \qquad i \neq j. \tag{7.29}$$

Proof Suppose the statement is true for $i, j \leqslant k$. Since $[d^0, \ldots, d^j]=[r^0, \ldots, r^j]$, by Lemma 7.1 we have in particular $r^k \cdot d^j=0$ for $j=0, \ldots, k-1$, so using (7.26)

$$r^{k+1} \cdot d^k=r^k \cdot d^j+\alpha_k\langle d^k, d^j \rangle=0, \qquad j=0, \ldots, k-1.$$

But α_k is optimal and we also have

$$r^{k+1} \cdot d^k=f'(\xi^k+\alpha_k d^k) \cdot d^k=\frac{d}{d\alpha}f(\xi^k+\alpha d^k)_{\alpha=\alpha_k}=0,$$

so $r^{k+1} \cdot d^j=0, j=0, \ldots, k$. Together with Lemma 7.1 this shows that $r^{r+1} \cdot r^j=0$ for $j=0, \ldots, k$, which proves (7.29) for $i, j \leqslant k+1$. Finally, to show (7.28) for $i, j \leqslant k+1$ we note that since $Ad^j \in [r^0, \ldots, r^{j+1}]$ by (7.26), we have by (7.29) that $\langle r^{k+1}, d^j \rangle=0, j=0, \ldots, k-1$. Together with (7.25c) and the induction hypothesis this proves that

$$\langle d^{k+1}, d^j \rangle = \langle -r^{k+1}, d^j \rangle + \beta_k \langle d^k, d^j \rangle = 0, \quad j=0, \ldots, k-1.$$

But we already know that $\langle d^{k+1}, d^k \rangle=0$ and thus we have proved (7.28) for $i, j \leqslant k+1$. The induction step is now complete and the lemma follows since the statement is clearly true for $i, j \leqslant 1$. □

We can now prove that the conjugate gradient method gives the exact solution after at most M steps:

Theorem 7.1 For some $m \leqslant M$, we have $A\xi^m=b$.

Proof By (7.29) the gradients $r^j, j=0, 1, \ldots$, are pairwise orthogonal and since there are in R^M at most M pairwise orthogonal non-zero vectors, it follows that $r^m=A\xi^m-b=0$ for some $m \leqslant M$. □

By Theorem 7.1 the conjugate gradient method gives, in the absence of round-off errors, the exact solution after at most M steps. In our applications, however, we will view the conjugate gradient method as an iterative method and the required number of iterations will be much smaller than M. To study the convergence properties of the method, we first note that by (7.25a) we have for $k=0, 1, \ldots$,

$$\xi^k-\xi^0=\sum_{j=0}^{k-1} \alpha_j d^j. \tag{7.30}$$

By the orthogonality (7.28) it follows that

$$\langle \xi^k, d^k \rangle = \langle \xi^0, d^k \rangle, \qquad k=0, 1, \ldots$$

Using also the fact that $A\xi=b$, we see that for $k=0, 1, \ldots$,

$$-r^k \cdot d^k = -(A\xi^k - A\xi) \cdot d^k = \langle \xi-\xi^k, d^k \rangle = \langle \xi-\xi^0, d^k \rangle,$$

which shows that (7.25b) can be written

$$\alpha_k = \frac{\langle \xi-\xi^0, d^k \rangle}{\langle d^k, d^k \rangle} \qquad k=0, 1, \ldots,$$

Thus, by (7.30) we have in particular

$$\langle \xi^k-\xi^0, d^j \rangle = \langle \xi-\xi^0, d^j \rangle \qquad j=0, 1, \ldots, k-1.$$

But this is the same as saying that $\xi^k-\xi^0$ is the projection of the initial error $\xi-\xi^0$ with respect to $<. , .>$ on the space

$$W_k=[d^0, \ldots, d^{k-1}]$$

spanned by the first k search directions, and thus

$$\|\xi-\xi^k\|_A \equiv \|\xi-\xi^0-(\xi^k-\xi^0)\|_A \leqslant \|\xi-\xi^0-\eta\|_A, \qquad \forall \eta \in W_k. \tag{7.31}$$

Recalling Lemma 7.1 and the fact that $r^0=A\xi^0-A\xi=-A(\xi-\xi^0)$, we see that

$$W_k=[r^0, Ar^0, \ldots, A^{k-1}r^0]=[A(\xi-\xi^0), \ldots, A^k(\xi-\xi^0)].$$

Using (7.31) we thus have the following result:

Theorem 7.2 For the conjugate gradient method (7.25),

$$\|\xi-\xi^k\|_A \leqslant \|p_k(A)(\xi-\xi^0)\|_A \leqslant \max_j |p_k(\lambda_j)| \; \|\xi-\xi_0\|_A, \quad \forall p_k \in \hat{P}_k,$$

where $\hat{P}_k$ is the set of polynomials $p_k(z)=\sum_{j=0}^{k} \beta_j z^j$, $\beta_j \in R$, of degree at most k with $\beta_0=1$.

To estimate the reduction of the initial error $\|\xi-\xi^0\|_A$ after k steps, it is by Theorem 7.2 sufficient to construct a polynomial p_k of degree at most k such that $p_k(0)=1$ and p_k is as small as possible on the interval $[\lambda_1, \lambda_M]$ containing the eigenvalues of A, i e, so that the quantity

$$\gamma_k = \max_{z \in [\lambda_1, \lambda_M]} |p_k(z)|$$

is as small as possible. The best polynomial is a *Chebyshev polynomial* well known in approximation theory, and the corresponding value of γ_k is (see e g [Ax]):

$$\gamma_k=2\left[\frac{\sqrt{k(A)}-1}{\sqrt{k(A)}+1}\right]^k, \quad k=0, 1, 2, \ldots .$$

Thus, for a given $\varepsilon>0$, to satisfy

$$\|\xi-\xi^k\|_A \leqslant \varepsilon \|\xi-\xi^0\|_A,$$

it is sufficient to choose n such that $\gamma_n \leqslant \varepsilon$, or by a simple computation, such that

$$n \geqslant \frac{1}{2}\sqrt{\varkappa(A)} \log \frac{2}{\varepsilon}. \tag{7.32}$$

We thus conclude that the required number of iterations for the conjugate gradient method is proportional to $\sqrt{\varkappa(A)}$ which should be compared with

$\varkappa(A)$ in the case of the gradient method. Thus, for $\varkappa(A)$ large, the conjugate gradient method is much more efficient than the gradient method. In a typical finite element application we have $\varkappa(A)=0(h^{-2})$, and thus in this case the required number of iterations would be of the order $0(h^{-1})$ for the conjugate gradient method and $0(h^{-2})$ for the gradient method.

Remark The subspace $W_k=[r^0, Ar^0, \ldots, A^{k-1}r^0]=[d^0, \ldots, d^{k-1}]$ is called the *Krylov subspace* related to the conjugate gradient method (7.25). By (7.31) we have that $\|\xi-\xi^k\|_A$ is the norm of the difference between the initial error $\xi-\xi^0$ and its projection on W_k. □

Example 7.2 Let us recall the simple minimization problem of Example 7.1,

$$(7.33) \qquad \min_{\eta\in R^2} \frac{1}{2}(\lambda_1\eta_1^2+\lambda_2\eta_2^2),$$

with $0<\lambda_1<<\lambda_2$. Applying the gradient method with optimal or constant step length, we have that the required number of iterations is proportional to λ_2/λ_1. Introducing the new variable $\zeta=(\zeta_1, \zeta_2)=(\sqrt{\lambda_1}\eta_1, \sqrt{\lambda_2}\eta_2)$, the problem takes the form

$$(7.34) \qquad \min_{\zeta\in R^2} \frac{1}{2}(\zeta_1^2+\zeta_2^2).$$

The condition number of the corresponding matrix is equal to 1, and the gradient method with optimal step length for (7.34) finds the exact solution $\zeta=0$ in just one iteration. This shows that a suitable change of variables may reduce the number of iterations significantly. We see that the very elongated elliptical level curves of (7.33) are replaced by the circular level curves of (7.34). The possibility of reducing the condition number for more general problems by a suitable change of variables corresponding to so-called preconditioning, will be discussed in Section 7.4 below. □

Problems

7.1 Show that β_k of (7.25d) can alternatively be computed as follows:

$$\beta_k=\frac{r^{k+1}\cdot r^{k+1}}{r^k\cdot r^k}.$$

7.2 Prove for the gradient method with optimal step length (7.8) that

$$\|\xi^{k+1}\|_A^2\leq\left(1-\frac{1}{\varkappa(A)}\right)\|\xi^k\|_A^2,$$

by proving

$$\frac{\|\xi^k\|_A^2-\|\xi^{k+1}\|_A^2}{\|\xi^k\|_A^2}=\frac{r^k\cdot r^k}{r^k\cdot Ar^k}\cdot\frac{r^k\cdot r^k}{r^k\cdot A^{-1}r^k}.$$

7.4 Preconditioning

We recall our quadratic minimization problem (7.1)

$$\text{(7.35)} \qquad \min_{\eta\in R^M} f(\eta)=\min_{\eta\in R^M}\left[\frac{1}{2}\eta\cdot A\eta-b\cdot\eta\right].$$

Let now E be a non-singular $M\times M$ matrix and introduce the new variable $\zeta=E\eta$ so that $\eta=E^{-1}\zeta$, and define

$$\tilde{f}(\zeta)=f(\eta)=f(E^{-1}\zeta)=\frac{1}{2}(E^{-1}\zeta)\cdot A(E^{-1}\zeta)-b\cdot E^{-1}\zeta=$$
$$=\frac{1}{2}\zeta\cdot E^{-T}AE^{-1}\zeta-E^{-T}b\cdot\zeta=\frac{1}{2}\zeta\cdot\tilde{A}\zeta-\tilde{b}\cdot\zeta,$$

where

$$\tilde{A}=E^{-T}AE^{-1},\ \tilde{b}=E^{-T}b,$$

and $E^{-T}=(E^{-1})^T$, where D^T denotes the transpose of the matrix D. Thus we can write the problem (7.35) using the new variable ζ as

$$\text{(7.36)} \qquad \min_{\zeta\in R^M}\left[\frac{1}{2}\zeta\tilde{A}\zeta-\tilde{b}\cdot\zeta\right].$$

The gradient method with constant steps α for this problem reads:

$$\text{(7.37)} \qquad \zeta^{k+1}=\zeta^k-\alpha(\tilde{A}\zeta^k-\tilde{b}).$$

The rate of convergence of this method depends on the condition number $\varkappa(\tilde{A})$. If $\varkappa(\tilde{A})<<\varkappa(A)$, then the gradient method for (7.36) will converge much faster than the same method applied to the original problem (7.35).

Before discussing how to choose the matrix E note that setting $\zeta=E\eta$ and multiplying by E^{-1} in (7.37), we get

$$\eta^{k+1}=\eta^k-\alpha E^{-1}E^{-T}(A\eta^k-b).$$

Thus, setting $C=E^TE$ so that $C^{-1}=E^{-1}E^{-T}$, we see that (7.37) corresponds to the following method for (7.35):

$$\text{(7.38)} \qquad \eta^{k+1}=\eta^k-\alpha C^{-1}(A\eta^k-b),\ k=0,1,\ldots,.$$

We say that this is a *preconditioned* version of the usual gradient method for (7.35) with the matrix C being the *preconditioner.* To compute η^{k+1} from (7.38) for a given η^k, we have to solve the system

$$Cd=(A\eta^k-b),$$

(note that we would not explicitly form C^{-1}).

We can now state the obviously desired properties of the matrix $C=E^TE$ (recall that $\tilde{A}=E^{-T}AE^{-1}$):

$$\text{(7.39a)} \qquad \varkappa(E^{-T}AE^{-1})<<\varkappa(A),$$

(7.39b) the system $Cd=e$ can be solved with few $(0(M))$ operations for a given right hand side e.

Suppose that $C=E^TE$ is the Cholesky factorization of C, and hence E is upper triangular. Then (7.39b) will be satisfied if E is essentially as sparse as A, i e, if the number of non-zero entries in each row of E is bounded independently of h. On the other hand, to satisfy (7.39a) the best choice would be $C=A=E^TE$ with E^TE the Cholesky factorization of A, in which case $\varkappa(E^{-T}AE^{-1})=1$. However, with this choice the matrix E is not as sparse as desired (cf the discussion of fill-in Section 6.4), and (7.39b) will be violated. With this background we are led to try to construct $C=E^TE$ such that E is sparse and E^TE is an approximate Cholesky factorization of A. We may require E to have a sparsity structure that is similar to that of A; for example we might allow an element e_{ij} of E to be non-zero only if the corresponding element a_{ij} of A is non-zero. To obtain an approximate factorization E^TE of A with this structure, we may perform a modified Gaussian elimination where non-zero elements appearing in the elimination process at "forbidden" locations are simply replaced by zeros. Such modified elimination processes (so-called *incomplete factorizations*) only take $0(M)$ operations and result in approximate factorizations with corresponding considerable reduction of the condition number, (e g, $\varkappa(E^{-T}AE^{-1})=0(h^{-1})$, see [Ax], [Me]).

7.5 Multigrid methods

Recently a class of methods for our typical system of equations (7.3) have been developed that are *optimal* in the sense that the required number of operations is of the order $0(M)$, where M is the number of unknowns (clearly $0(M)$ is optimal since this amount of work is required just to write down the solution). These methods are the so-called *multi-grid* methods (see e g [BD], [Bra], [Hac]). A multi-grid method is an iterative method where one uses a collection of successively coarser finite element grids.

To give an idea of the basic features of the multi-grid method we consider the standard finite element method of Section 1.4 on a triangulation T_h obtained by subdividing each triangle of a coarser triangulation T_{2h} into four triangles as in Fig 1.14. Let the correspongding finite element spaces be V_h and V_{2h}. Then the corresponding matrices A_h and A_{2h} have dimension $M\times M$ and $(M/4)\times(M/4)$, respectively. Assume that we want to solve the system $A_h\xi=b$, and to start with assume that the system $A_{2h}\eta=d$ for a given d can be solved in $0(M/4)$ operations. A step of the multigrid method leading from a given approximation $\xi^k\in R^M$ to an improved approximation $\xi^{k+1}\in R^M$ now consists of two substeps: a smoothing step and a coarse grid correction. The smoothing step consists of m usual gradient steps:

$$\eta^{i+1}=\eta^i-\alpha(A_h\eta^i-b), \qquad i=0, 1, \ldots, m-1, \tag{7.40}$$

with α suitably chosen (cf Section 7.2), and $\eta^0=\xi^k$. This step gives the approximation $\xi^{k+\frac{1}{2}}\equiv\eta^m$. The coarse grid correction is obtained as follows: Let $\delta\in V_{2h}$ be the solution of the problem

$$a(\delta, v)=(f, v)-a(u^{k+\frac{1}{2}}, v) \quad \forall v\in V_{2h}, \tag{7.41}$$

where $u^{k+\frac{1}{2}}\in V_h$ is the finite element function with nodal values $\xi^{k+\frac{1}{2}}$. Let $\delta^{k+\frac{1}{2}}\in R^{\frac{M}{4}}$ be the vector of nodal values of δ and define

$$\xi^{k+1}=\xi^{k+\frac{1}{2}}+\delta^{k+\frac{1}{2}},$$

where the components of $\delta^{k+\frac{1}{2}}\in R^M$ are given by the components of $\delta^{k+\frac{1}{2}}\in R^{\frac{M}{4}}$ for the nodes of T_{2h}, and the value of $\delta^{k+\frac{1}{2}}$ at other nodes are obained by linear interpolation from the values at the nodes of T_{2h}. We note that the correction step (7.41) corresponds to a problem of the form $A_{2h}\eta=d$ which can be solved in $0(M/4)$ operations by assumption.

To sum up, a multigrid step leading from ξ^k to ξ^{k+1} consists of a simple smoothing step together with a coarse grid correction requiring few operations. Under suitable assumptions one can prove that there is a constant C independent of ξ and k such that

$$|\xi-\xi^{k+1}|\leq\frac{C}{m}|\xi-\xi^k|,$$

which proves that for m sufficiently large each multigrid step reduces the error significantly.

The algorithm is now applied recursively so that to solve a problem of the form $A_{2h}\eta=d$ in the step described above, we invoke a coarser grid with corresponding matrix A_{4h}, assuming that T_{2h} is obtained as above by refinement of the corser grid T_{4h}. This gives a procedure where we work on a sequence $T_h, T_{2h}, T_{4h}, T_{8h}, \ldots$, of successively coarser grids ending with a coarsest grid for which the corresponding linear system can be solved by direct Gauss elimination with few operations. One can show that this combined process will give a solution of the original matrix problem $A_h\xi=b$ in $0(M)$ operations.

The reason why the multigrid method is so efficient is, roughly speaking, the following: In the smoothing step the high frequency components of the error (corresponding to large eigenvalues) are significantly reduced. This fact is easy to understand from the analysis of the gradient method in Section 7.2 above. Further, in the coarse grid correction the low and medium frequency components of the error are also significantly reduced and thus in each multigrid step all components of the error are reduced significantly.

7.6 Work estimates for direct and iterative methods

Here we collect the principal results presented above concerning the amount of work required to solve our typical system of equations

$$A\xi=b \tag{7.42}$$

by direct and iterative methods, where A is a sparse, symmetric and positive definite $M\times M$ matrix. We then suppose that (7.42) is related to a second order elliptic problem in R^d, $d=2$ or 3. In this case $M=0(h^{-d})$ and the condition number $\varkappa(A)=0(h^{-2})$. We further assume that in the preconditioned variants of the conjugate gradient method the condition number is reduced to $0(h^{-1})$. Also, in the Cholesky factorization we assume that A is stored as a band matrix with band width $0(h^{-d+1})$. With these assumptions we have an asymptotic work estimate for the solution of (7.42) of the form $0(M^{\alpha})$, where the exponents α are given by:

Method	dim 2	dim 3
Band-Cholesky: factorization	2	2.33
back-substitution	1.5	1.67
Nested dissection: factorization	1.5	2
back-substitution	1	1.33
Conjugate gradient	1.5	1.33
Preconditioned conjugate gradient	1.25	1.17
Multigrid	1	1

Fig 7.3

Clearly, the multigrid and preconditioned conjugate gradient method have the most favourable exponents and for M large enough will be superior to band-Cholesky and nested dissection. This holds particularly for $d=3$. However, the multigrid method requires a rather complex program with considerable overhead to organize the computations while band-Cholesky requires little overhead. Thus, for a given M it is not clear which method would require the least total cost, and of course this cost also depends on the problem and on the implementation of the particular method.

We also note that one sometimes wants to solve the system $A\xi=b$ many times with the same A but different right hand sides b. For example we may want to compute the stress distribution in an elastic body under various loads. In this case we may factorize the matrix A once and for all and then only a back-substitution will be required for each new right hand side. In such cases band-Cholesky becomes comparatively more competitive but still is asymptotically inferior to the preconditioned conjugate gradient and multigrid methods.

To sum up we may say that, roughly speaking, band-Cholesky may be used for coarse to medium fine discretizations in two dimensions whereas iterative methods multigrid or preconditioned conjugate gradient type would be advantageous for large three-dimensional problems and for very fine discretizations in two dimentions. Let us remark that these conclusions should be valid at least on well-structured problems with coefficients that are not varying too much and using e g quasi-uniform finite element meshes. For problems with highly variable coefficients and very complicated solutions it may be difficult to find iterative methods with good convergence properties and in such cases Gaussian elimination may be the only realistic alternative at present.

Remark 7.3 To store only the nonzero elements of a sparse symmetric $M\times M$ matrix $A=(a_{ij})$, one may use a vector $a=(a(i))$ containing the elements in the lower triangular part of A ordered row by row, together with a vector $ac=(ac(i))$, with $ac(i)$ the number of the column in A containing the element $a(i)$, and the vector $ad=(ad(j))$, with $ad(j)=i$ where $a(i)=a_{jj}$. As an example, if

$$A=\begin{bmatrix} a_{11} & & & & \\ a_{12} & a_{22} & & \text{sym} & \\ 0 & a_{32} & a_{33} & & \\ a_{41} & 0 & a_{43} & a_{44} & \\ a_{51} & 0 & a_{53} & 0 & a_{55} \end{bmatrix},$$

then we have

$$a=(a_{11}, a_{12}, a_{22}, a_{32}, a_{33}, a_{41}, a_{43}, a_{44}, a_{51}, a_{53}, a_{55}),$$
$$ac=(1, 1, 2, 2, 3, 1, 3, 4, 1, 3, 5),$$
$$ad=(1, 3, 5, 8, 11).$$

□

Problem

7.3 Determine the asymptotic work estimates corresponding to Fig 7.3 for a fourth order elliptic problem.

7.7 The condition number of the stiffness matrix

If A is the stiffness matrix related to an elliptic problem of order 2m, then the condition number $\varkappa(A)$ is under suitable conditions estimated by

(7.43) $\qquad \varkappa(A)=0(h^{-2m})$.

Let us prove this result in the standard case $m=1$, $A=(a_{ij})$, $a_{ij}=a(\varphi_i, \varphi_j)$,

$$a(v, w)=\int_\Omega \nabla v \cdot \nabla w \, dx,$$

with $\varphi_1, \ldots, \varphi_M$, the usual basis for $V_h=\{v\in H_0^1(\Omega): v|_K\in P_1(K), K\in T_h\}$, where $\Omega\subset R^2$. This is the case studied in Section 1.4.

We shall assume that the family $\{T_h\}$ of triangulations $T_h=\{K\}$ satisfies the following conditions: There are positive constants β_1 and β_2 independent of $h=\max_{K\in T_h} h_K$ such that for all $K\in T_h$, $T_h\in\{T_h\}$,

(7.44a) $\qquad h_K\geqslant\beta_1 h,$

(7.44b) $\qquad \dfrac{\varrho_K}{h_K}\geqslant\beta_2,$

where h_K and ϱ_K are defined as in Section 4.2. The condition (7.44a) states that all elements K of T_h are of roughly the same size. Such triangulations are said to be *quasi-uniform*.

We recall that the bilinear form $a(.\,,.)$ is $H_0^1(\Omega)$ – elliptic, i e, there is a positive constant α such that

(7.45) $\qquad a(v, v)\geqslant\alpha\|v\|^2_{H^1(\Omega)} \qquad\qquad \forall v\in H_0^1(\Omega).$

The estimate (7.43) with $m=1$ will easily follow from the following result:

Lemma 7.3 There are constants c and C only depending on the constants β_i in (7.44), such that for all $v=\sum_{i=1}^{M}\eta_i\varphi_i\in V_h$

(7.46) $\qquad ch^2|\eta|^2\leqslant\|v\|^2\leqslant Ch^2|\eta|^2,$

(7.47) $\qquad a(v, v)\equiv\int_\Omega|\nabla v|^2dx\leqslant Ch^{-2}\|v\|^2,$

where $\|v\|=\|v\|_{L_2(\Omega)}$.

Remark The estimate (7.47) is a so-called *inverse estimate;* here we estimate the L_2-norm of the gradient of v in terms of the L_2-norm of v itself. This is not possible for a general function v, but it is possible for the functions v in V_h at the price of the factor h^{-1}. □

We postpone the proof of Lemma 7.3 and show how to prove (7.43) using the lemma. We recall that if $v=\sum_{i=1}^{M}\eta_i\varphi_i$, then

$$a(v, v)=\eta\cdot A\eta,$$

so that by (7.46) and (7.47)

$$\frac{\eta\cdot A\eta}{|\eta|^2}=\frac{a(v, v)}{|\eta|^2}\leqslant Ch^{-2}\frac{\|v\|^2}{|\eta|^2}\leqslant C^2 \qquad \forall\eta\in R^M. \tag{7.48}$$

On the other hand, we have by (7.45) and (7.46) since trivially $\|v\|_{H^1(\Omega)}\geqslant\|v\|$,

$$\frac{\eta\cdot A\eta}{|\eta|^2}=\frac{a(v, v)}{|\eta|^2}\geqslant\alpha\frac{\|v\|^2}{|\eta|^2}\geqslant c\alpha h^2 \qquad \forall\eta\in R^M. \tag{7.49}$$

Together, (7.48) and (7.49) prove that there are constants c and C such that

$$\lambda_{max}\leqslant C, \quad \lambda_{min}\geqslant ch^2,$$

which gives the desired result $\varkappa(A)=\dfrac{\lambda_{max}}{\lambda_{min}}\leqslant Ch^{-2}$.

Remark 7.4 Note that it is natural to scale the matrix A, by multiplying with a constant of order $0(h^{-2})$, so that $\lambda_{max}=0(h^{-2})$ and $\lambda_{min}=0(1)$ (cf (1.25)). With this scaling A will be a discrete counterpart of the Laplace operator with eigenvalues ranging from 0(1) to $0(h^{-2})$ (recall that the eigenvalues of the Laplace operator on a bounded domain lie in the unbounded interval (Λ, ∞) for some positive Λ. □

Let us give

Proof of Lemma 7.3 It is sufficient to show that for each triangle $K\in T_h$ with vertices a^i and $v\in P_1(K)$, we have

$$ch_K^2\sum_{i=1}^{3}|v(a^i)|^2\leqslant\|v\|^2_{L^2(K)}\leqslant Ch_K^2\sum_{i=1}^{3}|v(a^i)|^2, \tag{7.50}$$

$$\int_K|\nabla v|^2dx\leqslant Ch_K^{-2}\int_K|v|^2dx, \tag{7.51}$$

with c and C independent of K and v. From these estimates the desired estimates (7.46) and (7.47) directly follow by summation over $K\in T_h$.

We first show that (7.50) and (7.51) hold when $K=\hat{K}$ where $\hat{K}$ is the *reference triangle* with vertices at (0, 0), (1, 0) and (0, 1) in a $(\hat{x}_1, \hat{x}_2)$ – plane (see Fig 7.3). Let $\hat{\lambda}_i$ be the usual basis functions of $P_1(\hat{K})$ and define

$$f_1(\hat{\eta})=\int_{\hat{K}}|\nabla\hat{v}|^2 d\hat{x},$$

$$f_2(\hat{\eta})=\int_{\hat{K}}\hat{v}^2 d\hat{x},$$

where $\hat{\eta}=(\hat{\eta}_1, \hat{\eta}_2, \hat{\eta}_3)$ and

$$\hat{v}(\hat{x})=\sum_{i=1}^{3}\hat{\eta}_i\hat{\lambda}_i(x), \qquad \hat{x}\in\hat{K}.$$

We observe that f_1 and f_2 are continuous functions of $\hat{\eta}\in R^3$. We now consider the quotient

$$f_3(\hat{\eta})=\frac{f_1(\hat{\eta})}{f_2(\hat{\eta})}, \qquad \hat{\eta}\in R^3, \quad \hat{\eta}\neq 0.$$

We want to prove that there is a constant C such that

$$(7.52) \qquad f_3(\hat{\eta})\leqslant C, \quad \hat{\eta}\in R^3, \quad \hat{\eta}\neq 0;$$

this inequality clearly corresponds to (7.51) in the case $K=\hat{K}$ since $h_{\hat{K}}=\sqrt{2}$. To prove (7.52) we first note that

$$f_3(\gamma\hat{\eta})=f_3(\hat{\eta}), \quad \forall\gamma\in R, \quad \gamma\neq 0,$$

i e, the function f_3 is *homogeneous of degree zero*. It is thus sufficient to prove that for some constant C

$$(7.53) \qquad f_3(\hat{\eta})\leqslant C \qquad \hat{\eta}\in B,$$

where $B=\{\hat{\eta}\in R^3: |\hat{\eta}|=1\}$. But f_3 is continuous on B (note in particular that $f_2(\hat{\eta})\neq 0$ for $\hat{\eta}\in B$) and B is a closed and bounded set in R^3, and thus f_3 has a maximum on B. This proves (7.53) and thus (7.52) and (7.51) follow in the case $K=\hat{K}$. In the same way we can prove (7.50) in this case.

It now remains to prove (7.50) and (7.51) for an arbitrary triangle $K\in T_h$. For simplicity assume that K is a triangle with vertices at (0, 0), (h, 0) and (0, h) so that $h_K=\sqrt{2}h$ and let the mapping $F: \hat{K}\rightarrow K$ be defined by (see Fig 7.3)

$$x=F(\hat{x})=(h\hat{x}_1, h\hat{x}_2), \quad \hat{x}\in\hat{K}.$$

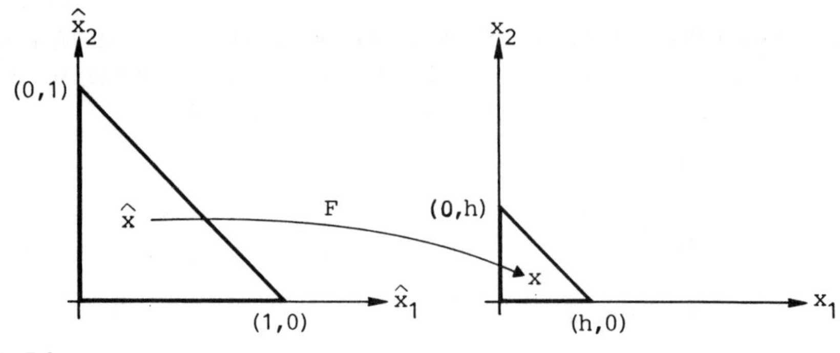

Fig 7.3

Given $v \in P_1(K)$ we now define

(7.54) $\quad \hat{v}(\hat{x}) = v(x) = v(F(\hat{x})), \quad \hat{x} \in \hat{K}.$

Clearly we then have $\hat{v} \in P_1(\hat{K})$. By the chain rule

$$\frac{\partial \hat{v}}{\partial \hat{x}_i} = \frac{\partial v}{\partial x_1}\frac{\partial x_1}{\partial \hat{x}_i} + \frac{\partial v}{\partial x_2}\frac{\partial x_2}{\partial \hat{x}_i} = \frac{\partial v}{\partial x_i}h \qquad i=1,\ 2,$$

and so $\nabla \hat{v} = h \nabla v$. Since $dx = h^2 d\hat{x}$, this gives

$$\int_K |\nabla v|^2 dx = \int_K h^{-2} |\nabla \hat{v}|^2 dx = \int_{\hat{K}} |\nabla \hat{v}|^2 d\hat{x}$$

$$\leqslant C \int_{\hat{K}} \hat{v}^2 d\hat{x} = C \int_K v^2 h^{-2} dx = Ch^{-2} \int_K v^2 dx,$$

where we used that (7.51) holds if $K = \hat{K}$. This proves (7.51) if K has vertices at (0, 0), (h, 0), (0, h), and in a similar way (7.50) can be shown in this case.

Finally, if K is an arbitrary triangle, then we introduce the linear mapping $F: \hat{K} \rightarrow K$ that maps $\hat{K}$ onto K:

$$x = F(\hat{x}) = a^1 + (a^2 - a^1)\hat{x}_1 + (a^3 - a^1)\hat{x}^2,$$

where a^i are the vertices of K. Arguing now as in the above special case and using the facts that $|a^i - a^1| \leqslant Ch_K$, $i=2, 3$, and $dx = Ch_K^2 d\hat{x}$ by (7.44), we now obtain (7.50) and (7.51) in the general case (cf Problem 7.4) and the proof is complete.

Remark The technique of working with a reference element $\hat{K}$ and linear mappings $F: \hat{K} \rightarrow K$ that map $\hat{K}$ onto the triangles $K \in T_h$, is very important also from a practical point of view. In this way it is often possible to use a simple transformation to obtain the stiffness matrix of a general element K from the stiffness matrix of the reference element $\hat{K}$. This is important since the direct computation of a stiffness matrix may require a not-negligible amount of work and the element stiffness matrices of all elements $K \in T_h$ have to be determined. □

Problems

7.4 Complete the proof of Lemma 7.3.

7.5 The condition (7.44a) stating that all elements have the same size can be relaxed. Prove without using (7.44a) that the estimate $\varkappa(A)=0(h^{-2})$ may be replaced by

$$\varkappa(A)=0(h_{min}^{-2}) \text{ where } h_{min}=\min_{K \in T_h} h_K.$$

7.6 Let $u_h \in V_h \subset L_2(\Omega)$ be the $L_2(\Omega)$-projection of $u \in L_2(\Omega)$ defined by (cf Problem 4.8),

$$(u_h, v)=(u, v) \qquad \forall v \in V_h.$$

In matrix form with the basis $\{\varphi_1, \ldots, \varphi_M\}$ for V_h, this problem takes the form $B\xi=b$, where $B=(b_{ij})$, $b_{ij}=(\varphi_i, \varphi_j)$, $b=(b_i)$, $b_i=(u, \varphi_i)$. Prove that B is positive definite. When V_h is piecewise polynomial on a quasi-uniform triangulation in R^2, show that $\varkappa(B)=0(1)$. The matrix B is called the *mass matrix* and will occur in Chapters 8 and 9.

8. FEM for parabolic problems

8.1 Introduction

In this chapter we give an introduction to finite element methods for linear parabolic problems. A typical such problem, modelling heat conduction in an isotropic body with heat capacity λ and conductivity μ and occupying a region $\Omega \subset R^d$, reads as follows:

$$
(8.1) \qquad
\begin{aligned}
\lambda \dot{u} - \operatorname{div}(\mu \nabla u) &= f && \text{in } \Omega \times I,\\
u &= 0 && \text{on } \Gamma_1 \times I,\\
\mu \frac{\partial u}{\partial n} &= 0 && \text{on } \Gamma_2 \times I,\\
u(x, 0) &= u^0(x) && x \in \Omega.
\end{aligned}
$$

Here $u(x, t)$ is the temperature at $x \in \Omega$ at time $t \in I = (0, T)$, where T is a given time, u^0 is a given initial temperature, f is a given heat production, Γ_1 and Γ_2 is a subdivision of the boundary Γ of Ω and $\dot{u} = \partial u / \partial t$. For simplicity we shall consider the following special case of (8.1) with $\lambda = \mu = 1$, $\Omega \subset R^2$ and $\Gamma_1 = \Gamma$:

$$
\begin{aligned}
&(8.2a) \qquad \dot{u} - \Delta u = f && \text{in } \Omega \times I,\\
&(8.2b) \qquad u = 0 && \text{on } \Gamma \times I,\\
&(8.2c) \qquad u(\cdot, 0) = u^0.
\end{aligned}
$$

Essential parts of the presentation based on (8.2) that follows, may directly be extended to the more general problem (8.1), (cf Example 2.7)

We will first consider a so-called *semi-discrete* analogue of (8.2) where we have discretized in space using the finite element method. To obtain a fully discrete problem we will then discretize time also. We shall see that the semi-discrete problem is an initial value problem for a system of ordinary differential equations. This will be a *stiff* system which will pose extra requirements on the stability of the methods to be used for the time-discretization (the notion of an initial value problem for a stiff system of ordinary differential equations will be explained in Section 8.3 below). For the time-discretization we shall first consider two classical methods for stiff problems: the *backward Euler method* and the *Crank-Nicolson method.* We shall then consider a recent method, the so-called *discontinuous Galerkin method,* based on using a finite element formulation in time with piecewise polynomials of degree q. In simple cases (e g $\lambda=\mu=1$ and $f=0$ in (8.1)) the discontinuous Galerkin method gives time-discretization methods which coincide with classical methods based on the so-called sub-diagonal Padé approximations. In particular, with $q=0$ one obtains the backward Euler method. The advantage of the discontinuous Galerkin method is that e g variable coefficients and non-zero right hand sides (and even non-linearities) present no complications in principle. Further, the fact that the method has a variational formulation is very useful in the analysis of the time-discretization error. In fact, one can derive precise error estimates for the discontinuous Galerkin method which make it possible to construct (for the first time) rational efficient methods for automatic time-step control which are of particular importance for stiff problems. We comment briefly on this topic in Section 8.4. The new possibilities offered by the discontinuous Galerkin method have been discovered only recently and are still under exploration, see [EJT], [J3], [EJ1], [JNT], [EJL]. For more information on finite element methods for parabolic problems, see also [Th], [LR] and the references therein.

8.2 A one-dimensional model problem

Before going into the discussion of the numerical methods for (8.2) we shall briefly indicate some of the main properties of the exact solution u of (8.2). For simplicity we will then consider the following one-dimensional model problem modelling heat conduction in a bar (cf (1.3)):

$$\frac{\partial u}{\partial t}-\frac{\partial^2 u}{\partial x^2}=f \qquad 0<x<\pi,\ t>0, \tag{8.3a}$$

$$u(0,t)=u(\pi,t)=0 \qquad t>0, \tag{8.3b}$$

$$u(x,0)=u^0(x) \qquad 0<x<\pi. \tag{8.3c}$$

In the case $f=0$, we have by separation of variables that the solution of (8.3) is given by

$$u(x,t)=\sum_{j=1}^{\infty} u_j^0 e^{-j^2 t}\sin(jx), \tag{8.4}$$

where

$$u_j^0=\sqrt{2/\pi}\int_0^\pi u^0(x)\sin(jx)dx, \qquad j=1, 2, \ldots,$$

are the Fourier coefficients of the initial data u^0 with respect to the orthonormal system $\{\sqrt{2/\pi}\sin(jx)\}_{j=1}^{\infty}$ in $L_2(0,\pi)$. By (8.4) we see that u(x,t) is a linear combination of sine waves sin (jx) with frequencies j and amplitudes $u_j^0 \exp(-j^2t)$. We may say that each component sin (jx) lives on a time scale of order $0(j^{-2})$ since $\exp(-j^2t)$ is very small for j^2t moderately large. In particular we have that high frequency components quickly get damped. Thus, the solution u(x,t) will become smoother as t increases. This of course fits with the intuitive idea of the nature of a diffusion process such as heat conduction. However, in general u(x,t) will not be smooth for t small, and we may have that $\|\dot{u}(t)\|=\|\dot{u}(\cdot,t)\|\to\infty$ as $t\to 0$, where $\|\cdot\|$ denotes the $L_2(0,\pi)$-norm. More precisely, the size of the derivates of u (with respect to t or x) for t small will depend on how quickly the Fourier coefficients u_j^0 decay with increasing j: For example, if $u^0(x)=\pi-x$ for $0<x<\pi$, then $u_j^0=C/j$, in which case $\|\dot{u}(t)\|\sim Ct^{-\alpha}$ with $\alpha=3/4$ as $t\to 0$, and if $u^0(x)$ is the "hat function" $u^0(x)=\min(x,\pi-x)$ for $0<x<\pi$, then $u_j^0=C/j^2$ in which case $\|\dot{u}(t)\|\sim Ct^{-\alpha}$ with $\alpha=1/4$ as $t\to 0$ (cf Problem 8.1). If u_j^0 decays faster than $j^{-2.5}$ as $j\to\infty$, then $\|\dot{u}(t)\|$ will be bounded as $t\to 0$, but higher derivatives may still be unbounded. In principle, the "smoother" the initial function u^0 is, the more rapidly u_j^0 decays as $j\to\infty$. Note that here a "smooth" initial function has to satisfy in particular the boundary conditions (8.3b).

An initial phase for t small where certain derivatives of u are large, is called an *initial transient*. Thus the exact solution of a parabolic problem in general will have an initial transient where certain derivatives are large, but the solution will become smoother as t increases. This fact is of importance when solving a parabolic problem numerically, since it is advantageous to vary the mesh size (in time and space) according to the smoothness of the exact solution u and thus use a fine mesh where u is non-smooth and increase the mesh size as u becomes smoother. Note that transients may also occur for $t>0$ if for example the right hand side f (or the boundary conditions) in (8.1)–(8.3) vary abruptly in time.

The basic stability estimates in our context for the problems (8.2) and (8.3) are in the case $f=0$:

$$\|u(t)\|\leqslant\|u^0\|, \qquad t\in I, \tag{8.5}$$

$$\|\dot{u}(t)\|\leqslant\frac{C}{t}\|u^0\|, \qquad t\in I. \tag{8.6}$$

For the problem (8.3) these estimates follow directly from (8.4) using Parseval's formula together with the facts that $0 \leq e^{-s} \leq 1$ and $0 \leq se^{-s} \leq C$ for $s \geq 0$. It is also possible to prove (8.5) and (8.6) using "energy methods" without relying on an explicit solution based on separation of variables (cf Problem 8.6 below). Note that (8.6) states that if $u^0 \in L_2(\Omega)$, then $\|\dot{u}(t)\| = 0(t^{-1})$ as $t \to 0$.

Let us now turn to the discussion of numerical methods for (8.2).

8.3 Semi-discretization in space

The semi-discrete analogue of (8.2) will be based on a variational formulation of (8.2) which we now describe. Letting $V = H_0^1(\Omega)$, multiplying (8.2a) for a given t by $v \in V$, integrating over Ω and using in the usual way Green's formula, we get with the notation of Section 1.4:

$$(\dot{u}(t),v) + a(u(t),v) = (f(t),v).$$

Thus, we are led to the following variational formulation of (8.2): Find $u(t) \in V$, $t \in I$, such that

$$(\dot{u}(t),v) + a(u(t),v) = (f(t),v) \qquad \forall v \in V,\ t \in I, \tag{8.7a}$$

$$u(0) = u^0. \tag{8.7b}$$

Now, let V_h be a finite-dimensional subspace of V with basis $\{\varphi_1, \ldots, \varphi_M\}$. For definiteness we shall assume that Ω is a polygonal convex domain and that V_h consists of piecewise linear functions on a quasi-uniform triangulation of Ω with mesh size h and satisfying the minimum angle condition (4.1). Replacing V by the finite-dimensional subspace V_h we get the following semi-discrete analogue of (8.7): Find $u_h(t) \in V_h$, $t \in I$, such that

$$(\dot{u}_h(t),v) + a(u_h(t),v) = (f(t),v) \qquad \forall v \in V_h,\ t \in I, \tag{8.8a}$$

$$(u_h(0),v) = (u^0,v) \qquad \forall v \in V_h. \tag{8.8b}$$

Let us rewrite (8.8) using the representation

$$u_h(t,x) = \sum_{i=1}^{M} \xi_i(t)\varphi_i(x), \qquad t \in I, \tag{8.9}$$

with the time-dependent coefficients $\xi_i(t) \in R$. Using (8.9) and taking $v = \varphi_j$, $j = 1, \ldots, M$, in (8.8), we get

$$\sum_{i=1}^{M} \xi_i(t)\,(\varphi_i, \varphi_j) + \sum_{i=1}^{M} \xi_i(t) a(\varphi_i, \varphi_j) = (f(t), \varphi_j), \quad j=1, \ldots, M, \quad t \in I,$$

$$\sum_{i=1}^{M} \xi_i(0)\,(\varphi_i, \varphi_j) = (u^0, \varphi_j) \qquad\qquad j=1, \ldots, M,$$

or in matrix form

(8.10a) $\quad B\dot{\xi}(t) + A\xi(t) = F(t), \quad t \in I,$

(8.10b) $\quad B\xi(0) = U^0,$

where $B=(b_{ij})$, $A=(a_{ij})$, $F=(F_i)$, $\xi=(\xi_i)$, $U^0=(U_i^0)$,

$$b_{ij} = (\varphi_i, \varphi_j) = \int_\Omega \varphi_i \varphi_j dx,$$

$$a_{ij} = a(\varphi_i, \varphi_j) = \int_\Omega \nabla \varphi_i \cdot \nabla \varphi_j \, dx,$$

$$F_i(t) = (f(t), \varphi_i), \; U_i^0 = (u^0, \varphi_i).$$

Recall that both the mass matrix B and the stiffness matrix A are symmetric and positive definite. Further $\varkappa(B)=0(1)$ and $\varkappa(A)=0(h^{-2})$ as $h \to 0$ (see Problem 7.6). Introducing the Cholesky decomposition $B=E^TE$ and the new variable $\eta=E\xi$, the problem (8.10) takes the slightly simpler form

$$(8.11) \qquad \begin{aligned} &\dot{\eta}(t) + \bar{A}\eta(t) = g(t), \quad t \in I, \\ &\eta(0) = \eta^0, \end{aligned}$$

where $\bar{A}=E^{-T}AE^{-1}$ is a positive definite symmetric matrix with $\varkappa(\bar{A})=0(h^{-2})$, $g=E^{-T}F$, $\eta^0=E^{-T}U^0$ and $E^{-T}=(E^{-1})^T=(E^T)^{-1}$. The solution of (8.11) is given by the following formula (see any book on ordinary differential equations):

$$(8.12) \qquad \eta(t) = e^{-\bar{A}t}\eta^0 + \int_0^t e^{-\bar{A}(t-s)} g(s) ds, \quad t \in I.$$

The problem (8.11) (and (8.10)) is an example of a stiff initial value problem, the stiffness being related to the fact that the eigenvalues of $\bar{A}$ are positive and vary considerably in size corresponding to $\varkappa(\bar{A})$ being large.

Let us now return to our semi-discrete problem in the formulation (8.8). A basic stability inequality for this problem, with for simplicity $f=0$, is obtained as follows: Taking $v=u_h(t)$ in (8.8a), we get

$$(\dot{u}_h(t), u_h(t)) + a(u_h(t), u_h(t)) = 0, \quad t \in I,$$

or with as above $\|\cdot\| = \|\cdot\|_{L_2(\Omega)}$,

$$\frac{1}{2}\frac{d}{dt}\|u_h(t)\|^2+a(u_h(t),\ u_h(t))=0,$$

so that recalling also (8.8b),

$$\|u_h(t)\|^2+2\int_0^t a(u_h(s),\ u_h(s))ds=\|u_h(0)\|^2\leqslant\|u^0\|^2,$$

and thus in particular,

(8.13) $\qquad \|u_h(t)\|\leqslant\|u_h(0)\|\leqslant\|u^0\|, \quad t\in I.$

This estimate is clearly analogous to the estimate (8.5) for the continuous problem. Note that (8.5) may also be proved in the same way as (8.13).

For the semi-discrete problem (8.8) one can prove the following almost optimal error estimate. Recall that we are assuming, for simplicity, that Ω is a convex polygonal domain and that V_h consists of piecewise linear functions on a quasi-uniform triangulation of Ω with mesh size h.

Theorem 8.1 There is a constant C such that if u is the solution of (8.2) and u_h satisfies (8.8), then

(8.14) $$\max_{t\in I}\|u(t)-u_h(t)\|\leqslant C\left(1+\left|\log\frac{T}{h^2}\right|\right)\max_{t\in I}\ h^2\|u(t)\|_{H^2(\Omega)}.$$

Proof The proof is based on a duality argument involving the following auxiliary problem: Given $t\in I$ let φ_h: $(0,\ t)\rightarrow V_h$ satisfy

(8.15a) $\qquad -(\dot{\varphi}_h(s),v)+a(\varphi_h(s),v)=0 \qquad \forall v\in V_h,\ s\in(0,\ t),$

(8.15b) $\qquad \varphi_h(t)=e_h(t),$

where $e_h(s)=u_h(s)-\tilde{u}_h(s)$ and $\tilde{u}_h(s)\in V_h$ satisfies

(8.16) $\qquad a(u(s)-\tilde{u}_h(s),v)=0 \qquad \forall v\in V_h,\ s\in(0,\ T).$

Now, taking $v=e_h(s)$ in (8.15a), using (8.7), (8.8), (8.16), writing $\theta(s)=u(s)-\tilde{u}_h(s)$, and integrating by parts in time we have

$$\|e_h(t)\|^2=\int_0^t[-(\dot{\varphi}_h(s),\ e_h(s))+a(\varphi_h(s),\ e_h(s))]ds+(\varphi_h(t),\ e_h(t))$$

$$=\int_0^t[(\dot{e}_h,\ \varphi_h)+a(e_h,\ \varphi_h)]ds+(\varphi_h(0),\ e_h(0))$$

$$=\int_0^t[(\dot{\theta},\ \varphi_h)+a(\theta,\ \varphi_h)]ds+(\theta(0),\ \varphi_h(0))$$

$$=-\int_0^t (\theta, \dot{\varphi}_h)ds+(\theta(t), \varphi_h(t)).$$

Thus, we have the following simple error representation formula

$$\|e_h(t)\|^2=-\int_0^t (\theta(s), \dot{\varphi}_h(s))ds+(\theta(t), \varphi_h(t)). \tag{8.17}$$

Now, (8.15) is equivalent to the ordinary differential equation (cf (8.10))

$$-B\dot{\zeta}(s)+A\zeta(s)=0, \qquad s\in(0, t),$$
$$\zeta(t)=\zeta^0.$$

Using the explicit solution of this problem corresponding to (8.12) (or (8.20) below) and using also Lemma 7.3, we easily find that there is a constant C independent of $e_h(t)$ and t, such that (cf Problem 8.2)

$$\|\varphi_h(s)\|\leqslant\|e_h(t)\|, \qquad 0\leqslant s\leqslant t, \tag{8.18}$$

$$\int_0^t \|\dot{\varphi}_h(s)\|ds\leqslant C(1+|\log \frac{t}{h^2}|)\|e_h(t)\|, \tag{8.19}$$

which combined with (8.17) proves that

$$\|e_h(t)\|\leqslant C(1+|\log \frac{t}{h^2}|) \max_{s\in(0, t)} \|\theta(s)\|.$$

Note that the estimates (8.18) and (8.19) correspond to the estimates (8.5) and (8.6) for the continuous problem. To complete the proof we now just note that $u-u_h=u-\tilde{u}_h+\tilde{u}_h-u_h=\theta-e_h$, and we then obtain the desired estimate (8.14) using the L_2 – estimate for $\theta(s)=u(s)-\tilde{u}_h(s)$ of Theorem 4.3. □

Remark Note that the constant C in (8.14) is in particular independent of T. □

8.4 Discretization in space and time

8.4.1 Background

We shall now consider some methods for time-discretization of the semi-discrete problem (8.8) resulting in fully discrete analogues of (8.7). Let us then first consider the related problem (8.11). The qualitative behaviour of the solution $\eta(t)$ of this stiff initial value problem is (of course) similar to that of the exact solution u of (8.2) as discussed in Section 8.2. In particular we have the following representation for the solution of (8.11) in the case g=0 (cf (8.4)):

$$\eta(t)=\sum_{j=1}^{M}(\eta^0, \chi^j)e^{-\mu_j t}\chi^j, \tag{8.20}$$

where $\{\chi^j\}_{j=1}^M$ are the orthonormal (in R^M) eigenvectors of $\bar{A}$ with corresponding eigenvalues $\mu_1\leqslant\ldots\leqslant\mu_M$ satisfying $\mu_1=0(1)$ and $\mu_M=0(h^{-2})$. Here the large eigenvalues μ_j correspond to rapidly "oscillating" eigenvectors χ^j while smaller eigenvalues correspond to "smoother" eigenvectors (cf Problem 8.1). By (8.20) we see that $\eta(t)$ has components that live on time scales in the wide range from $0(h^{-2})$ to $0(1)$, that high frequency components of $\eta(t)$ are quickly damped and that $\eta(t)$ in general will have an initial transient. Note that the stiffness of (8.11) is reflected by the fact that the solution $\eta(t)$ contains components with vastly different time scales.

As indicated, stiff problems like (8.11) put special demands on the methods to be used for time discretization. First, for stability reasons one has, in order to avoid excessively small time steps, to use so called *implicit* methods, i e, methods requiring the solution of a system equations at each time step. Secondly, one would like to use methods which automatically adapt the size of the time steps according to the smoothness of $\eta(t)$ and thus automatically take smaller time steps in a transient and larger steps when $\eta(t)$ becomes smoother.

We will first briefly consider two classical methods for time-discretization of (8.8) or equivalently (8.10) and then the more recent discontinuous Galerkin method together with methods for automatic time step control. Let $0=t_0<t_1<\ldots<t_N=T$ denote a subdivision of I and write $I_n=(t_{n-1}, t_n)$ and let $k_n=t_n-t_{n-1}$ be the local time step.

8.4.2 The backward Euler and Crank-Nicolson methods

In the classical *backward Euler* method for the semi-discrete problem (8.8) we seek approximations $u_h^n\in V_h$ of $u(\cdot, t_n)$, $n=0,\ldots, N$, satisfying

$$\left(\frac{u_h^n-u_h^{n-1}}{k_n}, v\right)+a(u_h^n,v) = (f(t_n), v) \qquad \forall v\in V_h,\ n=1,2,\ldots,N, \tag{8.21a}$$

$$(u_h^0,v)=(u^0,v) \qquad \forall v\in V_h. \tag{8.21b}$$

Clearly, (8.21a) has been obtained from (8.8a) by replacing the derivative $\dot{u}_h(t_n)$ by the difference quotient $(u_h^n-u_h^{n-1})/k_n$ with discretization error $0(k_n)$.

For a given u_h^{n-1} we have that (8.21a) corresponds to the following positive definite symmetric system of equations for the unknown u_h^n:

$$(8.22) \qquad (B+k_nA)\xi^n=B\xi^{n-1}+k_nF(t_n),$$

where

$$u_h^n=\sum_{i=1}^{M}\xi_i^n\varphi_i.$$

A basic stability estimate for (8.21), with f=0 for simplicity, is obtained by taking $v=u_h^n$ in (8.21a) to yield

$$\|u_h^n\|^2-(u_h^n,u_h^{n-1})+a(u_h^n,u_h^n)k_n=0.$$

Using here the fact that

$$(u_h^n,u_h^{n-1})\leqslant \frac{1}{2}\|u_h^n\|^2+\frac{1}{2}\|u_h^{n-1}\|^2,$$

we conclude that for n=1, . . ., N,

$$\frac{1}{2}\|u_h^n\|^2-\frac{1}{2}\|u_h^{n-1}\|^2+a(u_h^n,u_h^n)k_n\leqslant 0,$$

so that by summation

$$\|u_h^n\|^2+2\sum_{m=1}^{n}a(u_h^m,u_h^m)k_m\leqslant\|u_h^0\|^2\leqslant\|u^0\|^2,$$

and in particular

$$(8.23) \qquad \|u_h^n\|\leqslant\|u_h^0\|\leqslant\|u^0\| \quad \text{for } n=1,\ldots,N,$$

which is clearly analogous to (8.13).

The other classical time-discretization method for (8.8) is the *Crank-Nicolson method:* Find $u_h^n\in V_h$, n=0, . . ., N such that

$$(8.24a) \qquad \left(\frac{u_h^n-u_h^{n-1}}{k_n},v\right)+a\left(\frac{u_h^n+u_h^{n-1}}{2},v\right)=\left(\frac{f(t_n)+f(t_{n-1})}{2},v\right) \quad \forall v\in V_h,\ n=1,\ldots,N,$$

$$(8.24b) \qquad (u_h^0,v)=(u^0,v) \quad \forall v\in V_h.$$

Here, the difference quotient $(u_h^n-u_h^{n-1})/k_n$ replaces $(\dot u(t_n)+\dot u(t_{n-1}))/2$ and the corresponding discretization error is $0(k_n^2)$. This time we obtain the following system of equations on each time level:

$$(8.25) \qquad \left(B+\frac{k_n}{2}A\right)\xi^n=\left(B-\frac{k_n}{2}A\right)\xi^{n-1}+k_n(F(t_n)+F(t_{n-1}))/2.$$

By taking $v=(u_h^n+u_h^{n-1})/2$ in (8.24a) we easily obtain again the stability inequality (8.23) in case $f=0$.

For the problem (8.11) in matrix form the backward Euler and Crank-Nicolson method read: Find $\eta^n \in R^M$, $n=1, 2, \ldots, N$ such that for $n=1, \ldots, N$

$$\frac{\eta^n-\eta^{n-1}}{k_n}+\bar{A}\eta^n=g(t_n), \tag{8.26}$$

$$\frac{\eta^n-\eta^{n-1}}{k_n}+\frac{1}{2}\bar{A}(\eta^n+\eta^{n-1})=\frac{1}{2}(g(t_n)+g(t_{n-1})). \tag{8.27}$$

In the case $g=0$, (8.26) leads to the following matrix equation for η^n:

$$(I+k_n\bar{A})\eta^n=\eta^{n-1}. \tag{8.28}$$

With the notation of Chapter 7, we have

$$|(I+k_n\bar{A})^{-1}|=\max_{j=1,\ldots,M}\frac{1}{1+k_n\mu_j}<1, \tag{8.29}$$

since the eigenvalues μ_j of $\bar{A}$ are positive. From (8.28) and (8.29), we have that

$$|\eta^n|\leqslant|\eta^{n-1}|\leqslant\ldots\leqslant|\eta^0|, \qquad n=1, \ldots, N, \tag{8.30}$$

which is another way of stating (8.23). Similarly we have for the Crank-Nicolson method

$$(I+\frac{1}{2}k_n\bar{A})\ \eta^n=(I-\frac{1}{2}k_n\bar{A})\ \eta^{n-1}, \tag{8.31}$$

and

$$|(I+\frac{1}{2}k_n\bar{A})^{-1}(I-\frac{1}{2}k_n\bar{A})|=\max_j\frac{|1-\frac{1}{2}k_n\mu_j|}{1+\frac{1}{2}k_n\mu_j}<1,$$

which again implies (8.30).

Not all time-discretization methods for (8.11) (or (8.10)) satisfy stability estimates of the form (8.30) (or (8.23)) regardless of the size of the time steps k_n. As an example, let us consider the so-called *forward Euler method* for (8.11):

$$\frac{\eta^n-\eta^{n-1}}{k_n}+\bar{A}\eta^{n-1}=g(t_{n-1}), \tag{8.32}$$

or, with $g=0$,

$$\eta^n=(I-k_n\bar{A})\eta^{n-1}. \tag{8.33}$$

Here

$$|I-k_n\bar{A}|=\max_j |1-k_n\mu_j|=|1-k_n\mu_M|\leqslant 1$$

only if $k_n\mu_M\leqslant 2$, or, $k_n\leqslant 2/\mu_M=0(h^2)$ since $\mu_M=0(h^{-2})$. This means that for the forward Euler method (8.32) we can only guarantee the stability inequality $|\eta^{n+1}|\leqslant|\eta^n|$ if the time step k_n is sufficiently small, or more precisely if

$$k_n\leqslant Ch^2. \tag{8.34}$$

In other words, the forward Euler method (8.32) is *conditionally stable* under the condition (8.34). If $|I-k_n\bar{A}|\geqslant\alpha>1$ with α independent of n, then the forward Euler will be unstable and useless for computational purposes and thus the method can only be used under the condition (8.34). This condition is very restrictive requiring very small time steps k_n if h is only moderately small.

One way of phrasing the stability condition (8.34) for the Euler forward method for (8.32) is to say that k_n has always to be chosen so small that the fastest time scale is resolved. Of course this is a natural condition in the initial phase of a transient where the "fastest" solution components play a role, but not so outside this phase. In contrast to the forward Euler method, the backward Euler and Crank-Nicolson methods are both stable regardless of the size of the time steps k_n, i e, these methods are *unconditionally stable.* This is a very desirable property of a time-discretization method for a parabolic problem.

In the backward Euler and Crank-Nicolson method we need to solve a system of equations at each time step (see (8.22), (8.25), (8.28), (8.31)), i e, these methods are implicit, whereas for the forward Euler method the solution η^{n+1} is directly given by η^n without solving any system, (see (8.33)), i e, this method is *explicit.* Clearly an implicit method requires more work per time step as compared with an explicit method. Thus, on the one hand we have unconditionally stable implicit methods and on the other hand conditionally stable explicit methods. The more efficient methods for parabolic problems belong to the first class; the extra cost involved at each step for an implicit method is more than compensated for by the fact that larger time steps may be taken (outside very fast transients, where accuracy requires very small time steps).

8.4.3 The discontinuous Galerkin method

We shall now consider the discontinuous Galerkin method for (8.8) which is based on using a finite element formulation to discretize in the time variable. To formulate this method we introduce for a given non-negative integer q the space

$$W_{hk}=\{v: I\to V_h: v|_{I_n}\in \mathbf{P}_q(I_n),\ n=1,\ldots,N\},$$

where

$$\mathbf{P}_q(I_n)=\{v: I_n\to V_h: v(t)=\sum_{i=0}^{q} v_i t^i \text{ with } v_i\in V_h\},$$

ie, W_{hk} is the space of functions on I with values in V_h that on each time interval I_n vary as polynomials of degree at most q. Notice that the functions v in W_{hk} may be discontinuous in time at the discrete time levels t_n. To account for this we introduce the notation

$$v_+^n=\lim_{s\to 0^+} v(t_n+s),\quad v_-^n=\lim_{s\to 0^-} v(t_n+s),$$

$$[v^n]=v_+^n-v_-^n,$$

where $[v^n]$ is the jump of v at t_n.

The discontinuous Galerkin method for (8.8) can now be formulated as follows: Find $U\in W_{hk}$ such that

$$A(U, v)=L(v)\qquad \forall v\in W_{hk}, \tag{8.35}$$

where

$$A(w, v)=\sum_{n=1}^{N}\int_{I_n}((\dot{w}, v)+a(w, v))dt$$
$$+\sum_{n=2}^{N}([w^{n-1}], v_+^{n-1})+(w_+^0, v_+^0),$$

$$L(v)=\int_I (f, v)dt+(u^0, v_+^0).$$

Since $v\in W_{hk}$ varies independently on each subinterval I_n, we may alternatively formulate (8.35) as follows: For $n=1,\ldots,N$, given U_-^{n-1}, find $U\equiv U|_{I_n}\in \mathbf{P}_q(I_n)$ such that

$$\int_{I_n}((\dot{U}, v)+a(U, v))dt+(U_+^{n-1}, v_+^{n-1})=$$
$$\int_{I_n}(f, v)dt+(U_-^{n-1}, v_+^{n-1}),\qquad \forall v\in \mathbf{P}_q(I_n), \tag{8.36}$$

where $U_-^0=u^0$.

For $q=0$, using the notation $U^n \equiv U^n_- \equiv U^{n-1}_+$, (8.36) reduces to the following problem: For $n=1, \ldots, N$, find $U^n \in V_h$ such that

$$(8.37) \qquad (U^n - U^{n-1}, v) + k_n a(U^n, v) = \int_{I_n} (f, v)dt \qquad \forall v \in V_h,\ n=1, \ldots, N,$$

where $U^0 = u^0$. This is a simple modification of the backward Euler scheme (8.21) where the right hand side involves an average of f over I_n rather than the value of f at t_n.

For $q=1$ we have that (8.36) is equivalent to the following system with

with $U(t) = U_0 + \frac{t - t_{n-1}}{k_n} U_1,\ t \in I_n,\ U_i \in V_h,$

$$(8.38) \qquad \begin{aligned} &(U_0, v) + k_n a(U_0, v) + (U_1, v) + \frac{1}{2} k_n a(U_1, v) \\ &\qquad\qquad = (U^{n-1}_-, v) + \int_{I_n} (f(s), v)ds, \qquad \forall v \in V_h, \\ &\frac{1}{2} k_n a(U_0, v) + \frac{1}{2}(U_1, v) + \frac{1}{3} k_n a(U_1, v) \\ &\qquad\qquad = \frac{1}{k_n} \int_{I_n} (s - t_{n-1})(f(s), v)ds, \qquad \forall v \in V_h. \end{aligned}$$

8.4.4 Error estimates for fully discrete approximations and automatic time and space step control

We now give an error estimate for the discontinuous Galerkin method (8.35) in the case $q=0$, ie, the backward Euler scheme (8.37) with, as above, V_h piecewise linear on a quasiuniform triangulation satisfying (4.1). The proof is given in Remark 8.1 below.

Theorem 8.2 Suppose there is a constant γ such that the time steps k_n satisfy $k_{n-1} \leq \gamma k_n$, $n=2, \ldots, N$ and let U^n be the solution of (8.37). Then there is a constant C depending only on γ and the constant β in (4.1) such that for $n=1, \ldots, N$,

$$(8.41) \qquad \|u(t_n) - U^n\| \leq C\left(1 + \log\frac{t_n}{k_n}\right)^{1/2}\left(\max_{m \leq n} \int_{I_m} \|\dot{u}(s)\|ds + \max_{t \leq t_n} h^2 \|u(t)\|_{H^2(\Omega)}\right).$$

Absorbing the "almost bounded" logarithmic quantity in the constant C, and using the trivial fact that

$$\int_{I_n} \|\dot{u}(s)\|ds \leq k_n\|\dot{u}\|_{\infty, I_n},$$

where $\|v\|_{\infty, J}=\sup_{s\in J} \|v(s)\|$, we can write (8.41) alternatively as follows

$$(8.42) \qquad \max_{t\in I} \|u(t)-U(t)\| \leq C \left(\max_{n\leq N} k_n\|\dot{u}\|_{\infty, I_n} + \max_{t\in I} h^2\|u(t)\|_{H^2(\Omega)}\right).$$

Here of course the first term on the right hand side bounds the time discretization error and the second term the space discretization error.

Suppose now $\delta>0$ is a given *tolerance* and that we want the time discretization error in (8.37) to be bounded by δ for all $t\in I$ (cf Problem 8.7). By (8.42) we see that this will be the case if

$$(8.43) \qquad k_n\|\dot{u}\|_{\infty, I_n} \leq \frac{\delta}{C}, \quad n=1, \ldots, N,$$

with C the constant in (8.42). This gives a rule for choosing the local time step k_n according to the size of $\|\dot{u}\|_{\infty, I_n}$. Of course, $\|\dot{u}\|_{\infty, I_n}$ is not known in advance, but it seems plausible that one would have

$$(8.44) \qquad k_n\|\dot{u}\|_{\infty, I_n} \sim \|U^n-U^{n-1}\|,$$

and thus we are led to the following criterion for correct choice of time steps for (8.37) involving only the computed solution U:

$$(8.45) \qquad \|U^n-U^{n-1}\| \simeq \frac{\delta}{C}.$$

One may satisfy this criterion at each time step e g by trial and error as follows: With U^{n-1} given, a first guess of the next time step k_n is made (e g $k_n=k_{n-1}$) and a corresponding U^n is computed. If $\|U^n-U^{n-1}\|$ is sufficiently close to δ/C, then U^n is accepted and otherwise k_n is decreased or increased to make $\|U^n-U^{n-1}\|\simeq\delta/C$. Variants of this procedure are possible. For example, as initial guess for k_n one may choose $\delta k_{n-1}/(C\,\|U^{n-1}-U^{n-2}\|)$.

Under reasonable assumptions one can show that (8.44) is correct (except possibly for a few initial steps) and thus one can prove that if the computationally verifiable criterion (8.45) is satisfied, then the time discretization error in (8.37) will be bounded by δ. A typical behaviour of $\|\dot{u}(t)\|$ for t moderately small, is given by $\|\dot{u}(t)\|\sim Ct^{-\alpha}$, where $0<\alpha<1$ (cf Section 8.2). In such a case the method based on (8.45) will thus automatically choose the correct time step sequence $k_n=\delta t_n^{\alpha}/C$ (cf Example 8.1 below).

Returning to the space discretization error in (8.42) let us notice that in (8.37) it is possible to use different spaces V_h^n for the space discretization on different time intervals I_n. The error estimate (8.42) also holds in this case, if for example $h_n^2 \leq ck_n$, where c is a sufficiently small constant, with now the space discretization error bounded by

$$(8.46) \qquad C \max_{n \leq N} h_n^2 \max_{s \in I_n} \|u(s)\|_{H^2(\Omega)},$$

where h_n is the mesh size in V_h^n (see [EJL]).

For simplicity assume now that f=0 in (8.2). By our assumption that Ω is convex, we then have by (8.2a) and (4.27) that $\|u(t)\|_{H^2(\Omega)} \leq C\|\dot{u}(t)\|$. To also control the space discretization error to the tolerance δ, we are therefore led to choose h_n depending on n so that

$$(8.47) \qquad h_n^2 \|\dot{u}\|_{\infty, I_n} \simeq \frac{\delta}{C}.$$

Again we may estimate the unknown quantity $\|\dot{u}\|_{\infty, I_n}$ using (8.44).

For the method (8.38) one may, under the assumptions of Theorem 8.2, prove an error estimate of the form

$$(8.48) \qquad \max_{t \in I} \|u(t)-U(t)\| \leq C \max_{n} (k_n^2 \|u^{(2)}\|_{\infty, I_n} + h_n^2 \max_{s \in I_n} \|u(s)\|_{H^2(\Omega)}),$$

where $u^{(2)} = \frac{d^2u}{dt^2}$ and where C contains a logarithmic factor as above. We also have the estimate

$$(8.49) \qquad \max_{n} \|u(t_n)-U_-^n\| \leq C \max_{n} (k_n^3 \|\Delta u^{(2)}\|_{\infty, I_n} + h_n^2 \max_{s \in I_n} \|u(s)\|_{H^2(\Omega)}),$$

which shows that (8.38) in fact is third order accurate in time at the discrete time levels t_n.

Relying on (8.49), we are led to control the time discretization error in (8.38) as follows:

$$(8.50) \qquad k_n^3 \|\Delta u^{(2)}\|_{\infty, I_n} \simeq \frac{\delta}{C}.$$

Again the unknown quantity $\|\Delta u^{(2)}\|_{\infty, I_n}$ may be estimated using the computed solution U. The method (8.38) with time step control (8.50) will typically require much fewer time steps than (8.37) with time step control (8.45). Although (8.38) is more costly per step than (8.37), in general (8.38) is much more efficient than (8.37) (cf Example 8.1).

In this section we have briefly indicated some important aspects of the problem of time discretization of parabolic problems. Of particular interest here is the discontinuous Galerkin method for which almost optimal error estimates can be obtained. These may be used as a basis for the design of rational methods for automatic time step control. With classical methods and techniques this was not achieved. For more information on this topic, see [J3], [EJ1], [JNT], [EJL].

Remark 8.1 The proof of the error estimate (8.41) for the backward Euler method (8.37), i e, the discontinuous Galerkin method (8.36) with q=0, is analogous to the proof of Theorem 8.1 and is again based on a duality argument. First, we introduce the interpolant $\tilde{U} \in W_{hk}$ defined by

$$\int_{I_n} a(\tilde{U}-u, v)ds=0 \qquad \forall v \in W_{hk},\ n=1, \ldots, N, \tag{8.51}$$

i e,

$$\tilde{U}^n \equiv \tilde{U}|_{I_n} = \frac{1}{k_n} \int_{I_n} \tilde{u}_h(s)ds,$$

where $\tilde{u}_h(s) \in V_h$ is given by (8.16). Now, let $Z \in W_{hk}$ satisfy

$$A(w, Z)=(w_-^N, U^N-\tilde{U}^N) \qquad \forall w \in W_{hk}, \tag{8.52}$$

i e, Z is a backward Euler approximation of the solution z of the problem:

$$-\dot{z}-\Delta z=0 \qquad \text{in } \Omega \times I,$$

$$z(T)=U^N-\tilde{U}^N.$$

Now, taking $w=U-\tilde{U} \in W_{hk}$ in (8.52) and using the fact that

$$A(u, v)=L(v) \qquad \forall v \in W_{hk},$$

we get, writing $\theta=u-\tilde{U}$ and recalling (8.51),

$$\|U^N-\tilde{U}^N\|^2=A(U-\tilde{U}, Z)=A(u-\tilde{U}, Z)$$

$$= \sum_{n=1}^{N} \int_{I_n} ((\dot{\theta}, Z)+a(u-\tilde{U}, Z))ds+ \sum_{n=2}^{N} ([\theta^{n-1}], Z_+^{n-1})$$

$$+(\theta_+^0, Z_+^0)=- \sum_{n=1}^{N-1} (\theta_-^n, [Z^n])+(\theta_-^N, Z_-^N).$$

This gives the error representation formula

$$\|U^N-\tilde{U}^N\|^2=(\theta_-^N, Z_-^N)- \sum_{n=1}^{N-1} (\theta_-^n, [Z^n]). \tag{8.53}$$

Now, corresponding to (8.18) and (8.19), we have the following stability estimates for (8.52) (cf Problem 8.6):

$$(8.54) \qquad \|Z_-^n\| \leqslant \|U^N - \tilde{U}^N\|, \qquad n=1, \ldots, N,$$

$$(8.55) \qquad \sum_{n=1}^{N-1} \|[Z^n]\| \leqslant C\left(1+\log \frac{T}{k_N}\right)^{1/2} \|U^N - \tilde{U}^N\|.$$

Together with (8.53) these estimates prove that

$$\|u(t_N) - U^N\| \leqslant C \left(1+\log \frac{T}{k_N}\right)^{1/2} \max_{t \in I} \|u(t) - \tilde{U}(t)\|,$$

which proves (8.41) for $n=N$. Since N may be replaced by n for $n=1, \ldots, N$, we thus obtain (8.41) by estimating $\|u(t) - \tilde{U}(t)\|$ as in the proof of Theorem 8.1. □

Remark 8.2 The stability estimates (8.54) and (8.55) for the discrete auxiliary problem (8.52) used in the above proof, correspond to the estimates (8.5) and (8.6) for the continuous problem (8.7). In particular, the near optimality of the error estimate (8.41) is a result of the use of the strong stability estimate (8.55). Conventional (non-optimal) error estimates for (8.37) only use the weaker stability estimate (8.54). □

Remark 8.3 Note that the constant C in (8.41) in particular is independent of t_n. This means that we may compute over very long time intervals essentially without growth of the global error. This reflects the parabolic nature of our problem. □

Remark 8.4 If we apply the backward Euler ($i=1$), the Crank-Nicolson method ($i=2$) and the discontinuous Galerkin method with $q=1$ ($i=3$) to the scalar problem

$$(8.56) \qquad \begin{aligned} &\dot{\eta} + \lambda\eta = 0, \qquad t>0, \\ &\eta(0) = \eta^0, \end{aligned}$$

where $\lambda>0$, we get the following time stepping methods

$$(8.57) \qquad \xi^n = r_i(k_n\lambda)\xi^{n-1}, \qquad i=1, 2, 3,$$

where for $x>0$,

$$r_1(x) = \frac{1}{1+x}, \qquad r_2(x) = \frac{1-\frac{x}{2}}{1+\frac{x}{2}}, \qquad r_3(x) = \frac{1-\frac{x}{3}}{1+\frac{2}{3}x+\frac{1}{6}x^2}.$$

The relations (8.57) should be compared with the following relation satisfied by the exact solution of (8.56):

$$\eta(t_n)=e^{-k_n\lambda}\ \eta(t_{n-1}).$$

Here of course the $r_i(x)$ are rational approximations of the exponential e^{-x}; we have that

$$e^{-x}-r_i(x)=0(x^{i+1}) \text{ as } x\rightarrow 0,$$

corresponding to the fact that the order of the method i is i, i=1, 2, 3. We also have

$$|r_i(x)|\leqslant 1 \qquad \text{for } x\geqslant 0,\ i=1, 2, 3,$$

and

$$r_i(x)\rightarrow 0 \qquad \text{as } x\rightarrow\infty \text{ for } i=1 \text{ and } 3,$$

but

$$r_2(x)\rightarrow -1 \qquad \text{as } x\rightarrow\infty. \tag{8.58}$$

By (8.58) we have that the rational function $r_2(x)$ associated with the Crank-Nicolson method does not behave like the exponential e^{-x} for x large. This means that the Crank-Nicolson method is not suitable for use in time intervals where the exact solution is non-smooth, for example in initial transients, where components corresponding to large eigenvalues λ play a role. □

Example 8.1 In Fig 8.1 we give results obtained by applying (8.38) with variable spaces V_h^n to the problem (8.3) with f=0 and $u_0(x)=1$, $0<x<\pi$ in which case $||\dot{u}(t)||\sim Ct^{-\alpha}$ as $t\rightarrow 0$ with $\alpha=\frac{3}{4}$, cf Section 8.2. The space and time step control was monitored by computational forms of (8.47) and (8.50). The number of space steps was restricted to be of the form 2^m, m=0, 1, ... We see that the error $||e(t)||_{L_2(0,\pi)}$ is, up to a factor 2, constant in time and of the order $0.2\delta=0.001$. Also notice that the time and space steps vary considerably in time and that the total number of steps $N\sim 30$. This example is taken from [EJL].

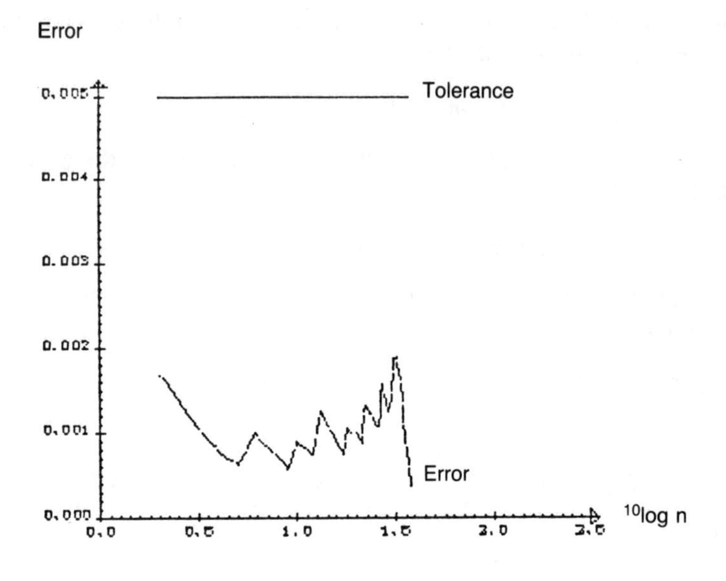

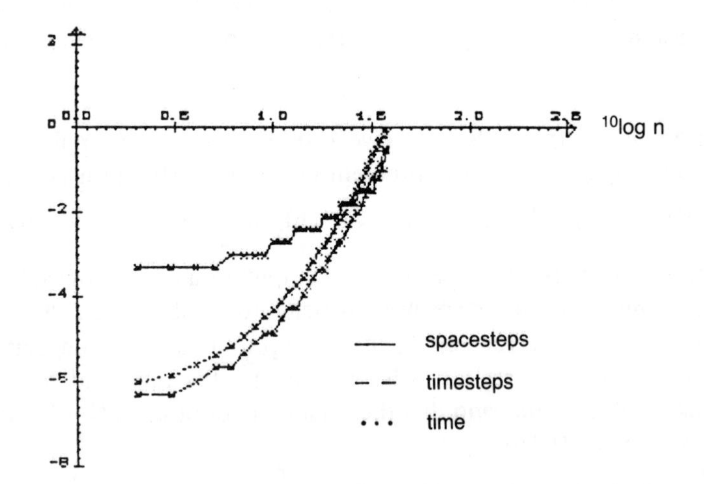

Fig 8.1 Fully discrete approximation of parabolic problem with automatic time and space step control

Problems

041

8.1 Consider the one-dimensional parabolic problem (8.3) with f=0.

(a) Prove using (8.4) that if $u^0(x)=\pi-x$, $0<x<\pi$, then

$$\|\dot{u}(t)\|\leqslant Ct^{-\frac{3}{4}} \qquad \text{as } t\to 0.$$

(b) Discretize in space using piecewise linear functions on a uniform partition with space step h and determine the corresponding ordinary differential equation (8.8). Determine the constant C in the stability requirement (8.34) for the forward Euler method with uniform step size k. Hint: The eigenvectors of the matrices B and A in this case are given by $v^j=(v_1^j, \ldots, v_M^j)$ with

$$v_i^j=\sin\left(\frac{ij\pi}{M+1}\right), \qquad M=\frac{1}{h}-1.$$

042

Also determine the condition numbers of the matrices B and A.

8.2 Let η be the solution of (8.11) with g=0 given by (8.20).

(a) Prove that

043

$$|\dot{\eta}(t)|+|\bar{A}\eta(t)|\leqslant\frac{C|\eta^0|}{t} \qquad t>0.$$

(b) Using (a) and the fact that $|\bar{A}\xi|\leqslant Ch^{-2}|\xi|$ prove that

$$\int_0^T(|\dot{\eta}(s)|+|\bar{A}\eta(s)|)ds\leqslant C\left(1+\left|\log\frac{T}{h^2}\right|\right)|\eta^0|.$$

8.3 Suppose the time steps k_n for (8.37) are chosen according to (8.43) with a non-smooth solution u satisfying $\|\dot{u}(t)\|\sim Ct^{-1+\varepsilon}$ for some $\varepsilon>0$. Compute the number of steps required if T=1 and compare with the number of steps required in the case of a smooth solution satisfying $\|\dot{u}(t)\|\sim C$.

)44

8.4 Propose an efficient method for solving the problem (8.22) related to the backward Euler method (8.21).

8.5 Compare computationally the methods (8.37) and (8.38) for the problem considered in Problem 8.1 with varying degree of smoothness of the initial data u_0. Use (8.45) for the time step control for (8.37) and (8.50) for the method (8.38). Compare with results obtained using a constant time step.

8.6 Prove for the method (8.37) with $f=0$ and under the assumption $k_n \leq Ck_{n-1}$, that

(a) $\|U^n\| \leq \|u^0\|$, $n=1, \ldots, N$,

(b) $\left(\sum_{n=1}^{N} t_n \left\| \frac{U^n - U^{n-1}}{k_n} \right\|^2 k_n \right)^{1/2} \leq C\|u^0\|$,

(c) $\sum_{n=1}^{N} \left\| \frac{U^n - U^{n-1}}{k_n} \right\| k_n \leq C\left(1+\log \frac{T}{k_1}\right)^{1/2} \|u^0\|$.

Also extend (a) to (8.35).

Hint: For (b) choose $v=t_n(U^n-U^{n-1})$ in (8.37). For (c) use (b) and Cauchy's inequality.

8.7 Using the error representation formula (8.17) prove the following variant of the estimate (8.14) for the semi-discrete problem (8.8) with $f=0$:

$$\|u(t)-u_h(t)\| \leq \frac{C}{t}\int_0^{t/2} \|\theta(s)\|ds + C\left(1+ \left|\log \frac{t}{h^2}\right|\right) \max_{\frac{t}{2}\leq s\leq t} \|\theta(s)\|$$

$$\leq C\left(1+\left|\log \frac{t}{h^2}\right|\right) \frac{h^2}{t} \|u^0\|.$$

This estimate shows that the error has optimal order for t bounded away from zero even for non-smooth initial data. In other words, to have a small error for t away from zero it is (for linear problems) not necessary to resolve an initial transient completely. Similar results hold for (8.30) and (8.38) (cf [Th]).

8.8 Consider the following time-dependent variant of the convection-diffusion problem (2.23):

$$\frac{\partial u}{\partial t} - \mu\Delta u + \beta_1 \frac{\partial u}{\partial x_1} + \beta_2 \frac{\partial u}{\partial x_2} = f \text{ in } \Omega\times I,$$

$$u=0 \text{ on } \Gamma\times I,$$

$$u(x,0)=u^0, \quad x\in\Omega.$$

Extend the methods (8.8), (8.21), (8.24), (8.35) to this problem and prove in the case $f=0$ a stability inequality analogous to (8.23).

9. Hyperbolic problems

9.1 Introduction

In previous chapters we have seen that the finite element method applied to linear elliptic and parabolic problems produces numerical methods with very satisfactory properties. We now turn to problems with mainly hyperbolic character, such as e g convection-diffusion problems with small or vanishing diffusion. Problems of this form typically occur in fluid mechanics, gas dynamics or wave propagation.

It was observed early on that, in contrast to what is the case for elliptic and parabolic problems, standard applications of the finite element method to hyperbolic problems lead to numerical schemes which frequently do not give reasonable results. More precisely, it was observed that standard finite element methods for hyperbolic problems do not work well in cases where the exact solution is not smooth. If the exact solution has e g a jump discontinuity, then the finite element solution will in general exhibit large spurious oscillations even far from the jump and will then not be close to the exact solution anywhere. This is of particular concern since in many interesting hyperbolic equations, the exact solution is not smooth. Only recently has it been possible to overcome these difficulties and construct modified non-standard finite element methods for hyperbolic problems with satisfactory convergence properties. In this chapter we will present these new finite element methods, the *streamline diffusion* (cf Remark 9.9) and *discontinuous Galerkin* methods, and compare them with standard methods. These new methods apply to first order hyperbolic problems such as e g convection-diffusion problems with small diffusion. We will also briefly discuss standard finite element methods for a second order hyperbolic problem, the wave equation for the Laplace operator. In this case improved finite element methods are still to be discovered.

9.2 A convection-diffusion problem

As an example of a linear hyperbolic-type equation let us consider the following convection-diffusion equation:

$$\frac{\partial u}{\partial t}+\mathrm{div}(u\beta)+\sigma u-\varepsilon\Delta u=0 \qquad \text{in } \Omega\times I. \tag{9.1}$$

Here u is a scalar unknown representing a concentration for example, $\beta=(\beta_1, \ldots, \beta_d)$ is a given velocity field, σ an absorption coefficient, $\varepsilon\geq 0$ a diffusion coefficient, $\Omega\subset R^d$ and $I=(0, T)$ is a given time interval. The equation (9.1) is of mixed hyperbolic-parabolic type with more or less hyperbolic or parabolic character depending on the size of ε and β. We assume here that ε is small, which means that (9.1) has mainly hyperbolic nature (if ε is not small then the material in Chapter 8 applies, cf Problem 8.8). In particular, with $\varepsilon=0$ we have the following purely hyperbolic equation:

$$\frac{\partial u}{\partial t}+\mathrm{div}(u\beta)+\sigma u=0 \quad \text{in } \Omega\times I, \tag{9.2}$$

or equivalently

$$\frac{\partial u}{\partial t}+\beta\cdot\nabla u+\gamma u+0 \quad \text{in } \Omega\times I, \tag{9.3}$$

where $\gamma=\sigma+\mathrm{div}\,\beta$. Let us briefly study this purely hyperbolic equation and first consider a stationary situation with u and β independent of time t, i e, let us consider the following equation

$$\beta\cdot\nabla u+\gamma u=0 \qquad \text{in } \Omega, \tag{9.4}$$

where $\beta=\beta(x)$ and $\gamma=\gamma(x)$ are given coefficients. The *streamlines* corresponding to the given velocity field $\beta=(\beta_1, \ldots, \beta_d)$ are given by the curves x(s), $x=(x_1, \ldots, x_d)$, where x(s) is a solution of the following system of ordinary differential equations:

$$\frac{dx_i}{ds}=\beta_i(x) \qquad i=1, \ldots, d.$$

These curves, parametrized by the parameter s, are also called the *characteristic curves* (or *characteristics*) of the problem (9.4). Assuming that β is Lipschitz-continuous (i e $|\beta(x)-\beta(y)|\leq C|x-y|\ \forall x, y\in\Omega$, for some constant C), there is for a given point $\bar{x}\in\Omega$ exactly one characteristic x(s) passing through $\bar{x}$, i e, there exists a unique function x(s) such that (cf Fig 9.1)

$$\frac{dx_i}{ds}=\beta_i(x), \quad i=1, \ldots, d, \qquad x(0)=\bar{x}. \tag{9.5}$$

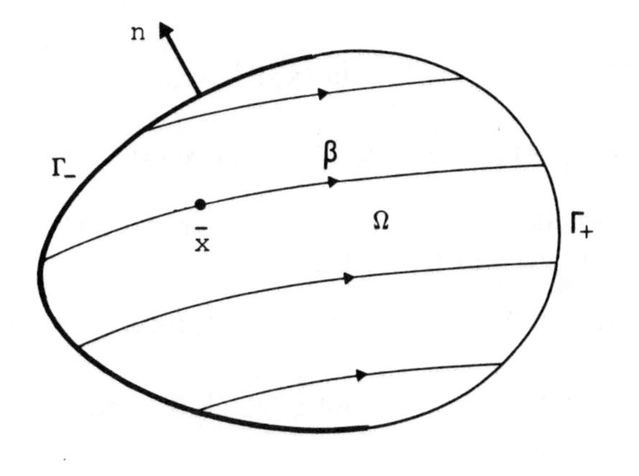

Fig 9.1

We now observe that if x(s) is a characteristic, then by the chain-rule we have

$$\frac{d}{ds}(u(x(s)))=\sum_{i=1}^{d}\frac{\partial u}{\partial x_i}\frac{dx_i}{ds}=\sum_{i=1}^{d}\frac{\partial u}{\partial x_i}\beta_i=\beta\cdot\nabla u,$$

so that by (9.4)

$$\frac{d}{ds}u(x(s))+\gamma u(x(s))=0. \tag{9.6}$$

Thus, along each characteristic the partial differential equation (9.4) is reduced to an ordinary differential equation. If the concentration u is known at one point on a given chaacteristic x(s), then u can be determined at other points on x(s) by integration of (9.6). As an example let us assume that u is given on the *inflow boundary* Γ_- defined by

$$\Gamma_-=\{x\in\Gamma:\ n(x)\cdot\beta(x)<0\},$$

where Γ is the boundary of Ω and n(x) is the outward unit normal to Γ at $x\in\Gamma$ (cf Fig 9.1). The concentration u at an arbitrary point $\bar{x}$ in Ω can then be determined by integration along the characteristic passing through $\bar{x}$ starting on Γ_-. In particular this means that in the problem (9.4), effects are propagated precisely along the characteristics.

It is important to notice that a solution u of (9.4) may be discontinuous across a characteristic. For instance, if the given concentration u on Γ_- has a jump discontinuity at some point $\bar{x}\in\Gamma$, then the solution u will be discontinuous across the entire characteristic passing through $\bar{x}$. As a simple example let us consider the following problem in R^2:

$$\frac{\partial u}{\partial x_1}=0 \qquad \text{in } \{x\in R^2{:}0<x_i<1\},$$

$$u(0, x_2)=1 \quad \text{for } 0<x_2<\frac{1}{2}, \qquad u(0, x_2)=0 \quad \text{for } \frac{1}{2}<x_2<1,$$

corresponding to taking $\beta_1\equiv1$, $\beta_2\equiv0$ and $\gamma=0$ in (9.4). Clearly the solution to this problem is given by (cf Fig 9.2).

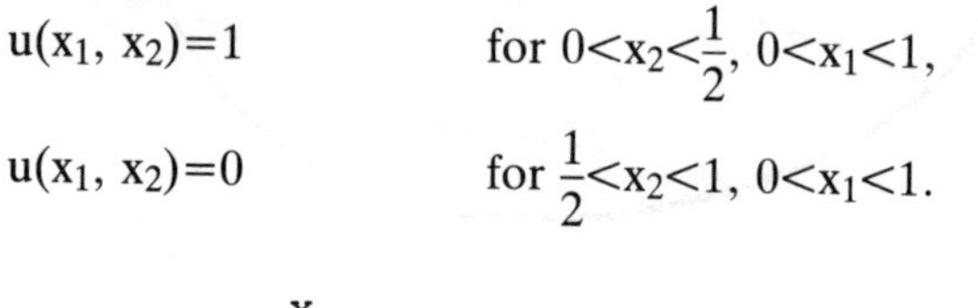

$$u(x_1, x_2)=1 \qquad \text{for } 0<x_2<\frac{1}{2}, \; 0<x_1<1,$$

$$u(x_1, x_2)=0 \qquad \text{for } \frac{1}{2}<x_2<1, \; 0<x_1<1.$$

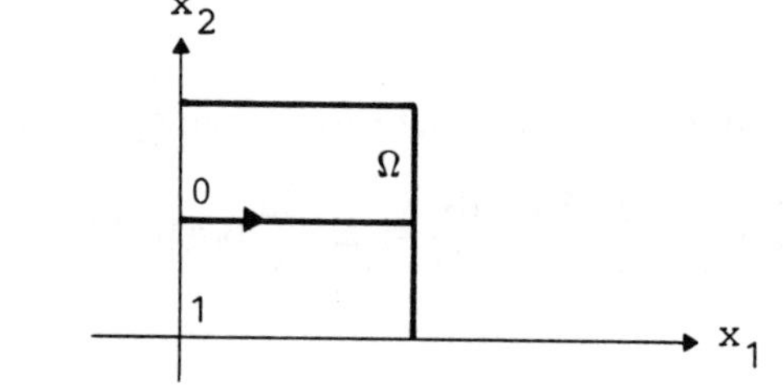

Fig 9.2

Let us now return to the time-dependent problem (9.3). If we here replace t by x_0 and let $\beta_0\equiv1$, then this equation can be written:

$$\text{(9.7)} \qquad \sum_{i=0}^{n} \beta_i \frac{\partial u}{\partial x_i}+\gamma u=0.$$

This is formally an equation of the same type as (9.4) and the discussion concerning (9.4) also applies to the equation (9.7). In particular, the characteristics of (9.7) are the curves $(x(t), t)$ in space-time, where $x(t)$ satsifies

$$\frac{dx_i}{dt}=\beta_i(x,t) \qquad i=1, \ldots, d,$$

(here the parameter s is chosen to be equal to $t=x_0$ corresponding to the equation $\frac{dx_0}{ds}=\beta_0=1$).

Remark 9.1 Another equation of the same form as (9.2), although nonlinear since here the velocity is unknown, is the continuity equation (or principle of conservation of mass) in gas dynamics:

$$\frac{\partial \varrho}{\partial t}+\operatorname{div}(\varrho v)=0,$$

where ϱ is a density and v a velocity. This equation, and additional equations expressing conservation of momentum and energy and a constitutive relation, constitutes a system of nonlinear hyperbolic equations which are the basic equations of gas dynamics. □

We will also consider briefly the following generalization of the scalar linear hyperbolic equation (9.3):

$$\frac{\partial u}{\partial t}+\sum_{j=1}^{d} A_j \frac{\partial u}{\partial x_j}+Bu=f, \tag{9.8}$$

where the A_j and B are m×m matrices depending on x and t, the A_j being symmetric, and u is an m-vector. We say that (9.8) is a linear *symmetric hyperbolic system.* In particular we shall consider a (positive) *Friedrichs' system* which is an equation of the form (9.8), together with certain boundary and initial conditions, satisfying a positivity condition. A simple example in one space dimension of a system of the form (9.8) is given by

$$\frac{\partial u}{\partial t}+\begin{bmatrix} 0 & -1 \\ -1 & 0 \end{bmatrix}\frac{\partial u}{\partial x}=0. \qquad u=\begin{bmatrix} u_1 \\ u_2 \end{bmatrix},$$

which is another way of writing the wave equation

$$\frac{\partial^2 w}{\partial t^2}-\frac{\partial^2 w}{\partial x^2}=0$$

using the notation $u_1=\frac{\partial w}{\partial t}$, $u_2=\frac{\partial w}{\partial x}$.

In the case of one space dimension (i e with d=1), a system of the form (9.8) together with appropriate boundary and initial conditions can be solved using the method of characteristics. In this case there are m characteristics $(x^i(t), t)$ $i=1, \ldots, m$, through each point $\bar{x}$, satisfying the equations

$$\frac{dx^i}{dt}=\lambda_i(x^i, t) \qquad i=1, \ldots, m,$$

where the $\lambda_i(x, t)$, $i=1, \ldots, m$, are the eigenvalues of the matrix $A_1(x, t)$.

9.3 General remarks on numerical methods for hyperbolic equations

Common methods for the numerical solution of hyperbolic equations are of the following types:

– method of characteristics,
– finite difference methods,
– finite element methods.

The method of characteristics may be used for scalar linear equations in several space dimensions and for linear systems in one space dimension. In practice this would come down to solving first the equations for the characteristics and then integrating along the characteristics, in both cases using some numerical method for integrating ordinary differential equations. In principle this is a very good method but it may not be so easy to use in practice, particularly not for a system. Further, for a mixed hyperbolic – parabolic equation like (9.1), this method cannot easily be used. For these more general problems one usually uses finite difference or finite element methods based on a *fixed mesh,* i e, a finite difference or finite element mesh that is not adapted to fit the characteristics of the particular problem to be solved. The use of a fixed mesh gives methods which are easy to program but it also may cause numerical difficulties if the exact solution is non-smooth with e g a jump discontinuity across a characteristic. In such a case conventional finite difference or finite element methods will produce approximate solutions which either oscillate (as standard Galerkin or centered finite difference methods) or excessively smear out a sharp front (as do classical artificial diffusion methods), see Fig 9.3.

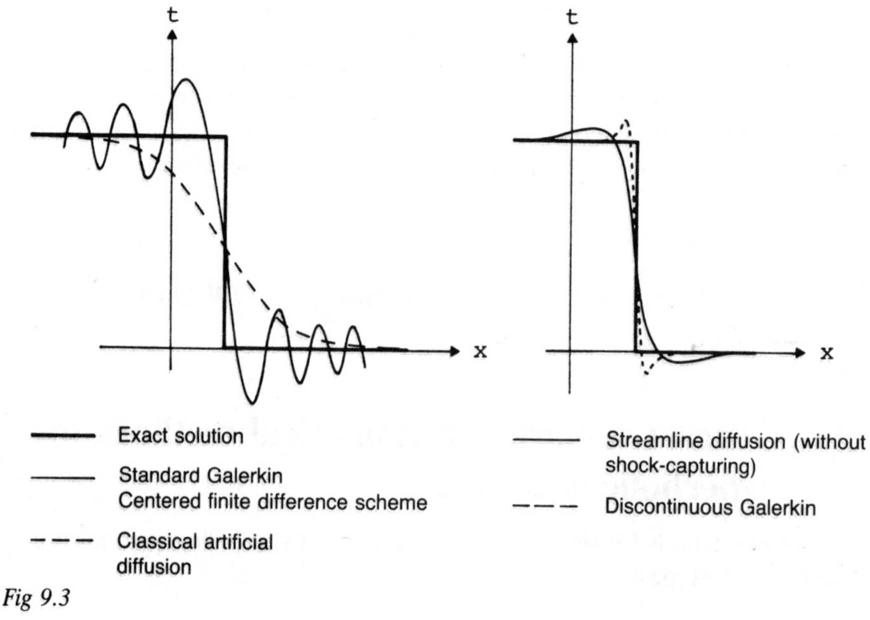

Fig 9.3

Thus, conventional methods are lacking in either stability or accuracy. Below we shall present the recently introduced streamline diffusion method and the discontinuous Galerkin method which at the same time have high order accuracy and good stability properties and which perform considerably better than the conventional methods (cf Fig 9.3 and Examples 9.2, 9.5 and 9.6 below).

9.4 Outline and preliminaries

We will first consider scalar problems of the form

$$(9.9)\qquad \begin{aligned} &\frac{\partial u}{\partial t}+\operatorname{div}(\beta u)+\sigma u-\varepsilon\Delta u=f \qquad (x,t)\in\Omega\times I,\\ &u(x,0)=u_0(x) \qquad x\in\Omega, \end{aligned}$$

with the stationary analogue

$$(9.10)\qquad \operatorname{div}(\beta u)+\sigma u-\varepsilon\Delta u=f \text{ in } \Omega,$$

together with boundary conditions, where Ω is a bounded domain in R^d, $I=(0,T)$ is a time interval, and the coefficients σ, $\varepsilon\geqslant 0$ and $\beta=(\beta_1,\ldots,\beta_d)$ depend smoothly on (x,t) or x. We assume that

$$(9.11)\qquad \frac{1}{2}\operatorname{div}\beta+\sigma\geqslant\alpha \qquad \text{in } \Omega\times I,$$

where $\alpha\geqslant 0$ is a constant with $\alpha>0$ in the stationary case. This condition will ensure the stability of the problems (9.9) and (9.10) for all $\varepsilon\geqslant 0$ (for ε small (9.11) can be relaxed, see e g [Na]). The boundary conditions may be of Dirichlet, Neumann or Robin (third) type. For simplicity we will consider two model problems with constant coefficients and Dirichlet boundary conditions, one stationary and one time-dependent. We leave the straight forward extension to variable coefficients and other boundary conditions to the problem section.

We shall consider the following finite element methods:

A. Standard Galerkin
B. Classical artificial diffusion
C. Streamline diffusion
D. Discontinuous Galerkin
E. Time discontinuous streamline diffusion.

The methods A, B and C apply to stationary mixed elliptic hyperbolic equations of the form (9.10) with ε small while method D is designed for purely hyperbolic problems of the form (9.9) and (9.10) with $\varepsilon=0$. Finally, E is intended for the time-dependent problem (9.9) with $\varepsilon>0$.

To conclude the chapter we shall discuss the application of the above methods to the case of Friedrichs' system and also consider some methods for second order hyperbolic equations, such as the wave equation.

We now state the two model problems to be discussed below. Let then Ω be a bounded convex polygonal domain in R^2 with boundary Γ and let $\beta=(\beta_1,\beta_2)$ be a constant vector with $|\beta|=1$. We shall consider the following stationary boundary value problem:

$$\begin{aligned} -\varepsilon\triangle u+u_\beta+u&=f \text{ in } \Omega, \\ u&=g \text{ on } \Gamma, \end{aligned} \tag{9.12}$$

where ε is a positive constant, and $v_\beta=\beta\cdot\nabla v$ denotes the derivative in the β-direction. The corresponding *reduced problem* obtained by setting $\varepsilon=0$ reads:

$$\begin{aligned} u_\beta+u&=f \text{ in } \Omega, \\ u&=g \text{ on } \Gamma_-, \end{aligned} \tag{9.13}$$

where Γ_- is the inflow boundary defined by

$$\Gamma_-=\{x\in\Gamma\colon n(x)\cdot\beta<0\},$$

where $n(x)$ is the outward unit normal to Γ at the point $x\in\Gamma$. The characteristics of the reduced problem (9.13) are straight lines parallel to β (cf Fig 9.4). We notice that in the reduced problem the boundary values are prescribed only on the inflow part Γ_-.

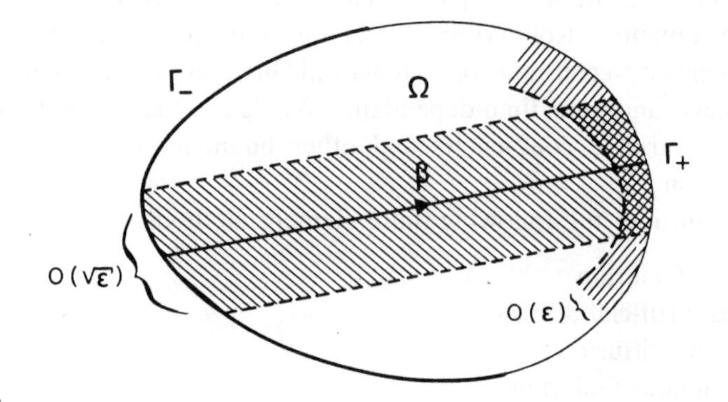

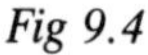
Fig 9.4

Let us briefly recall some basic facts concerning the regularity of the exact solutions u of (9.12) and (9.13). As already noted, the solution of the reduced problem (9.13) may be discontinuous with a jump across a characteristic if the boundary data g is discontinuous, for example. In the full problem (9.12) with $\varepsilon>0$, the solution is continuous in Ω, and such a jump will be "spread out" on a region of width $0(\sqrt{\varepsilon})$ around the characteristic. Such a narrow region (for ε small) where u (or some derivative of u) rapidly changes, is called a *layer.* If the values attained by the solution u of the reduced problem on the *outflow boundary* $\Gamma_+=\Gamma\setminus\Gamma_-$ do not coincide with the boundary value g specified in the full problem, then the solution of the latter problem will have a *boundary layer* at Γ_+. The thickness of this layer will be $0(\varepsilon)$, cf Fig 9.4.

Let $J=(0, 1)$ be a space interval, $I=(0, T)$ a time interval, and $\Omega=J\times I$. Then the time-dependent model problem is as follows:

$$
\begin{aligned}
&u_t+u_x-\varepsilon u_{xx}=f && \text{in } \Omega,\\
(9.14)\qquad &u(x,0)=u_0(x) && x\in J,\\
&u(x,t)=g(x,t) && x=0,1,\ t\in I,
\end{aligned}
$$

with the corresponding reduced problem:

$$
\begin{aligned}
&u_t+u_x=f && \text{in } \Omega,\\
(9.15)\qquad &u(x,0)=u_0(x) && x\in J,\\
&u(0,t)=g(t) && t\in I.
\end{aligned}
$$

Clearly the problem (9.15) has (except for the u-term) the same form as (9.13). The characteristics of (9.15) are straight lines in the (x, t)-plane with direction (1, 1) and the inflow boundary is given by the points (x, t) with $x=0$ or $t=0$ (cf Fig 9.5).

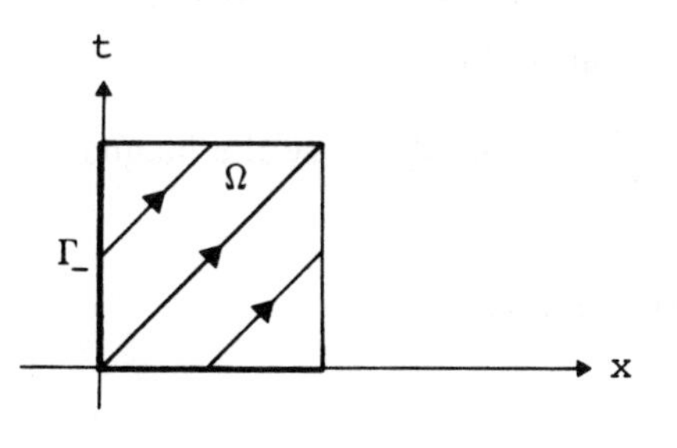

Fig 9.5

We use the following notation when discussing the methods A–D for the stationary problems (9.12) and (9.13):

$$(v, w)=\int_\Omega vw\,dx,\ (\nabla v,\nabla w)=\int_\Omega \nabla v\cdot\nabla w\,dx,$$

$$\|v\|=\|v\|_{L_2(\Omega)},\ \|v\|_s=\|v\|_{H^s(\Omega)},$$

$$\langle v, w\rangle=\int_\Gamma vw\ n\cdot\beta\ ds,$$

$$\langle v, w\rangle_-=\int_{\Gamma_-} vw\ n\cdot\beta\ ds, \qquad \langle v, w\rangle_+=\int_{\Gamma_+} vw\ n\cdot\beta\ ds,$$

$$|v|=(\int_\Gamma v^2|n\cdot\beta|ds)^{1/2},$$

where $\Gamma_+=\Gamma\setminus\Gamma_-=\{x\in\Gamma:n(x)\cdot\beta\geq 0\}$. We notice that by Green's formula

$$(v_\beta, w)=\langle v, w\rangle-(v, w_\beta).$$

Further let $\{T_h\}$ be a family of, for simplicity, quasi-uniform triangulations $T_h=\{K\}$ of Ω with mesh size h which satisfy as usual the minimum angle condition. For a given positive integer r we introduce the finite element space

$$V_h=\{v\in H^1(\Omega):\ v|_K\in P_r(K)\ \forall K\in T_h\},$$

ie, V_h is the space of continuous piecewise polynomial functions of degree r. From the approximation theory of Chapter 4, we have that for any $u\in H^{r+1}(\Omega)$ there exists an interpolant $\tilde{u}^h\in V_h$ such that

(9.16a) $\qquad \|u-\tilde{u}^h\|\leq Ch^{r+1}\|u\|_{r+1}.$

(9.16b) $\qquad \|u-\tilde{u}^h\|_1\leq Ch^r\|u\|_{r+1}.$

Moreover, if the derivatives of u of order r+1 are bounded on $\bar{\Omega}$, then

$$|u-\tilde{u}^h|\leq Ch^{r+1},$$

and with somewhat less stringent regularity requirements (see [Ci])

(9.16c) $\qquad |u-\tilde{u}^h|\leq Ch^{r+1/2}\|u\|_{r+1}.$

In the proofs below we will often use the inequality

$$2ab\leq\varepsilon a^2+\varepsilon^{-1}b^2,$$

for a, b real numbers and $\varepsilon>0$.

9.5 Standard Galerkin

Let us first consider the standard Galerkin method for the stationary model problem (9.12) with $\varepsilon>0$ and let us then for simplicity assume that the boundary data g is zero. This problem can be given the following variational formulation: Find $u \in H_0^1(\Omega)$ such that

$$(9.17) \qquad \varepsilon(\nabla u, \nabla v) + (u_\beta + u, v) = (f, v) \qquad \forall v \in H_0^1(\Omega).$$

Let now the finite element space

$$\mathring{V}_h = \{v \in V_h : v = 0 \text{ on } \Gamma\}$$

be given. The standard Galerkin method for (9.17) reads: Find $u^h \in \mathring{V}_h$ such that

$$(9.18) \qquad \varepsilon(\nabla u^h, \nabla v) + (u_\beta^h + u^h, v) = (f, v) \qquad \forall v \in \mathring{V}_h.$$

This method will perform well if $\varepsilon \geqslant h$, but if $\varepsilon << h$ then this method may produce an oscillating solution which is not close to the exact solution. To get an idea of what may happen, let us consider the following simple one-dimensional example:

Example 9.1 Consider the boundary value problem

$$(9.19) \qquad \begin{aligned} &-\varepsilon u_{xx} + u_x = 0, \ 0 < x < 1, \\ &u(0) = 1, \ u(1) = 0, \end{aligned}$$

with $0 < \varepsilon << 1$. The solution of this problem is given by

$$u(x) = a(1 - e^{-\frac{1-x}{\varepsilon}}), \ a = (1 - e^{-\frac{1}{\varepsilon}})^{-1}.$$

For ε small $u(x)$ is close to 1 except in a layer at $x=1$ of width $0(\varepsilon)$ where u decays from 1 to 0, see Fig 9.6.

If we apply the standard Galerkin method with piecewise linear functions on a uniform mesh with mesh length h to (9.19), we obtain the following system of equations for the values U_i of the finite element approximation u_h at the gridpoints $x_i = ih$, $i = 0, 1, \ldots, N$, where $x_N = 1$:

$$(9.20) \qquad \begin{aligned} &-\frac{\varepsilon}{h^2}[U_{i+1} - 2U_i + U_{i-1}] + \frac{1}{2h}[U_{i+1} - U_{i-1}] = 0, \ i = 1, \ldots, N-1, \\ &U_0 = 1, \ U_N = 0. \end{aligned}$$

We notice that (9.20) may also be viewed as a difference scheme with a *central* difference approximation $(U_{i+1} - U_{i-1})/2h$ for the convective term u_x. Now, if N is odd and ε is very small, then the solution U_i of (9.20) is approximately equal to 1 for i even and equal to 0 for i odd, and we get a solution that oscillates in the whole region and that is not close to the exact solution (cf Fig 9.6). □

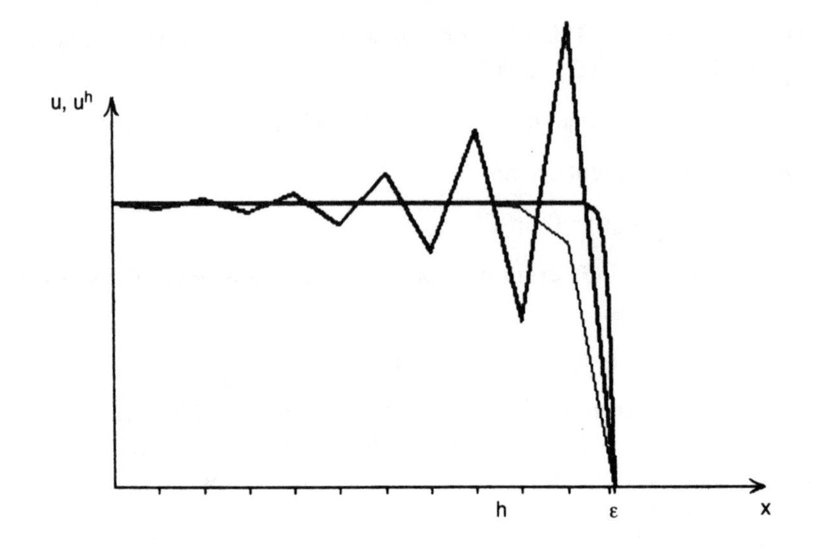

Fig 9.6 Exact solution and approximate solution by the standard Galerkin method (solid lines) for (9.19) with ε=0.01, h=1/11, (cf also Problem 9.5)

To sum up, the standard Galerkin method (9.18) may produce an oscillating solution if $\varepsilon<h$ and the exact solution is non-smooth. However, if the exact solution happens to be smooth, then the standard Galerkin method will produce good results even if $\varepsilon<h$ (cf Problem 9.2).

Let us now turn to the reduced problem (9.13) with $\varepsilon=0$. We will consider two variants of the standard Galerkin method.

Standard Galerkin with strongly imposed boundary conditions:
Find $u^h \in V_h$ with $u^h=g$ at the nodes on Γ_- such that

$$(9.21) \qquad (u^h_\beta+u^h, v)=(f, v) \qquad \forall v \in V_h \text{ with } v=0 \text{ on } \Gamma_-.$$

Standard Galerkin with weakly imposed boundary conditions:
Find $u^h \in V_h$ such that

$$(9.22) \qquad (u^h_\beta+u^h, v)-\langle u^h, v\rangle_-=(f,v)-\langle g,v\rangle_- \qquad \forall v \in V_h.$$

Let us analyze the method (9.22) (we leave the method (9.21) to Problem 9.3).

Introducing the notation

$$b(w, v)=(w_\beta+w, v)-<w, v>_-,$$

$$l(v)=(f,v)-<g,v>_-,$$

this methods reads: Find $u_h \in V_h$ such that

$$(9.23) \qquad b(u^h, v)=l(v) \qquad \forall v \in V_h.$$

Since the exact solution u satisfies (9.13), we clearly also have

$$b(u, v)=l(v) \qquad \forall v \in V_h,$$

and thus by subtraction we get the following equation for the error $e=u-u_h$:

$$(9.24) \qquad b(e, v)=0 \qquad \forall v \in V_h.$$

The stability of the method (9.22) is a consequence of the following property of the bilinear form b:

Lemma 9.1 For any $v \in H^1(\Omega)$ we have

$$b(v,v)=||v||^2+\frac{1}{2}|v|^2.$$

Proof By Green's formula

$$(v_\beta,v)=-(v,v_\beta)+<v,v>,$$

so that

$$(v_\beta,v)=\frac{1}{2}<v,v>=\frac{1}{2}<v,v>_+ +\frac{1}{2}<v,v>_-.$$

Hence

$$b(v,v)=||v||^2+\frac{1}{2}<v,v>_+ +\frac{1}{2}<v,v>_- - <v,v>_-=$$

$$=||v||^2+\frac{1}{2}<v,v>_+ -\frac{1}{2}<v,v>_- = ||v||^2+\frac{1}{2}|v|^2,$$

since $n\cdot\beta\geq 0$ on Γ_+ and $n\cdot\beta<0$ on Γ_-. □

Since (9.22) is equivalent to a linear system of equations with as many unknowns as equations, we obtain uniqueness and hence also existence of a solution from Lemma 9.1. Let us now prove an error estimate for the standard Galerkin method (9.22).

Theorem 9.1 There is a constant C such that if u satisfies (9.13) and $u^h \in V_h$ is the solution of (9.22), then

$$\|u-u^h\|+|u-u^h| \leq Ch^r\|u\|_{r+1}. \tag{9.25}$$

Proof Let $\tilde{u}^h \in V_h$ be the interpolant of u satisfying (9.16) and write $\eta^h=u-\tilde{u}^h$ and $e^h=u^h-\tilde{u}^h$ so that $e^h=\eta^h-e$, where $e=u-u^h$. By Lemma 9.1 and (9.24) with $v=e^h \in V_h$, we have

$$\|e^h\|^2+\frac{1}{2}|e^h|^2=b(e^h, e^h)=b(\eta^h, e^h)-b(e, e^h),$$

$$=b(\eta^h, e^h)=(\eta^h_\beta, e^h)+(\eta^h, e^h)-<\eta^h, e^h>_-$$

$$\leq\|\eta^h_\beta\|^2+\|\eta\|^2+\frac{1}{2}\|e^h\|^2+|\eta^h|^2+\frac{1}{4}|e^h|^2.$$

Recalling (9.16) we have that

$$\|\eta^h_\beta\|+\|\eta^h\|+|\eta^h|\leq Ch^r\|u\|_{r+1},$$

and thus

$$\|e^h\|+|e^h|\leq Ch^r\|u\|_{r+1}.$$

Since $e=e^h-\eta_h$, the desired inequality now follows from the triangle inequality and the proof is complete. □

The estimate (9.25) proves that if the exact solution u of the reduced problem (9.13) happens to be smooth so that $\|u\|_{r+1}$ is finite, then the standard Galerkin method (9.22) will converge at the rate $0(h^r)$. Although this rate is one power of h from being optimal, it shows that the standard Galerkin method will perform rather satisfactorily in this case. However, in general u will not be smooth and in this case the standard Galerkin method gives poor results (the error estimate (9.25) is useless, for example, if u is discontinuous since then $\|u\|_1=\infty$).

Problems

9.1 For r=1, 2, estimate the norm $\|u\|_r$ of the solution u of the one-dimensional problem (9.19) in terms of ε.

9.2 Prove an error estimate for the standard Galerkin method (9.18). Explain why this estimate does not necessarily imply that this method performs well for ε small.

9.3 Prove an error estimate for the method (9.21).

9.4 Consider for $\varepsilon>0$ the problem

$$-\varepsilon\Delta u+\mathrm{div}(\beta u)+\sigma u=f \quad \text{in } \Omega,$$
$$u=0 \quad \text{in } \Gamma,$$

with variable coefficients $\beta(x)$ and $\sigma(x)$ satisfying the condition (9.11) with $\alpha>0$. Formulate the standard Galerkin method for this problem and prove a stability estimate.

9.6 Classical artificial diffusion

The simplest way to handle the difficulties connected with the standard Galerkin method (9.18) with $\varepsilon<h$ and (9.21)–(9.22) is to avoid these situations completely. This can be done either by decreasing h until $\varepsilon>h$, which may impractical if ε is very small, or simply by solving, instead of the original problem with diffusion term $-\varepsilon\Delta u$, a modified problem with diffusion term $-h\Delta u$ obtained by adding the term $-\delta\Delta u$, where $\delta=h-\varepsilon$. This is the idea of the classical *artificial diffusion* (or viscosity) *method*. To be precise, this method for solving (9.12) with $\varepsilon<h$ reads: Find $u^h\in \mathring{V}_h$ such that

$$\text{(9.26)} \qquad h(\nabla u^h,\nabla v)+(u^h_\beta+u^h,v)=(f,v) \qquad \forall v\in \mathring{V}_h.$$

This method produces non-oscillating solutions but has the drawback of introducing a considerable amount of extra diffusion. In particular, this method introduces a diffusion term $-hu_{\eta\eta}$ acting in the direction η perpendicular to the streamlines ("crosswind" diffusion), and a sharp front or jump across a streamline will be considerably smeared out. Moreover, due to the added term $-\delta\Delta u$ such a method is at most first order accurate, and the error is at best $O(h)$ even for smooth solutions.

9.7 The streamline diffusion method

It turns out that to considerably reduce the oscillations in the standard Galerkin method (9.18) in the case $\varepsilon<h$, it is sufficient to add a term $-\delta u_{\beta\beta}$ where $\delta=h-\varepsilon$, ie, a diffusion term acting only in the direction of the

streamlines. Such a modified artificial diffusion method would read: Find $u^h \in \mathring{V}_h$ such that

$$(9.27) \qquad \varepsilon(\nabla u^h, \nabla v)+\delta(u^h_\beta, v_\beta)+(u^h_\beta+u_h, v)=(f,v) \quad \forall v \in \mathring{V}_h,$$

where $\delta=h-\varepsilon$. This method introduces less crosswind diffusion than the classical artificial diffusion method (9.26), but still corresponds to an $0(h)$-perturbation of the solution of the original problem.

However, it is possible to introduce the magic term $\delta(u^h_\beta, v_\beta)$ appearing in (9.27) without such a perturbation. Let us first see how this may be done in the case $\varepsilon=0$.

9.7.1 The streamline diffusion method with $\varepsilon=0$

Let us start from the standard Galerkin method (9.22) with weakly imposed boundary conditions. If, in the terms (. , .), we replace the test function $v \in V_h$ by $v+hv_\beta$, we get the streamline diffusion method: Find $u^h \in V_h$ such that

$$(9.28) \qquad (u^h_\beta+u^h, v+hv_\beta)-(1+h) \langle u^h, v\rangle_- \\ =(f,v+hv_\beta)-(1+h) \langle g,v\rangle_- \qquad \forall v \in V_h,$$

where for convenience we have also multiplied the boundary terms by the factor $(1+h)$. We notice the presence of the term $h(u^h_\beta, v_\beta)$. Further, we notice that the relation (9.28) is valid if we replace u^h by the solution u of (9.13), i e, the method (9.28) is consistent with (9.13) and does not introduce an $0(h)$-perturbation as do (9.26) and (9.27).

Let us now analyze the method (9.28) and introduce the following notation

$$B(w,v)=(w_\beta+w,v+hv_\beta)-(1+h) \langle w,v\rangle_-,$$
$$L(v)=(f,v+hv_\beta)-(1+h) \langle g,v\rangle_-.$$

The method (9.28) can then be formulated as follows: Find $u^h \in V_h$ such that

$$B(u^h,v)=L(v) \qquad \forall v \in V_h.$$

Moreover, as we have just noted, the exact solution of (9.13) satisfies

$$B(u,v)=L(v) \qquad \forall v \in H^1(\Omega),$$

and by subtraction we thus have the following error equation:

$$(9.29) \qquad B(e,v)=0 \qquad \forall v \in V_h,$$

where $e=u-u^h$.

We will prove an error estimate in the following norm

$$||v||_\beta=(h||v_\beta||^2+||v||^2+\frac{1+h}{2}\ |v|^2)^{1/2}.$$

This choice of norm is related to the following stability property of the bilinear form B(. , .).

Lemma 9.2 For any $v\in H^1(\Omega)$ we have

$$B(v,v)=||v||_\beta.$$

Proof By Green's formula we have

$$(v_\beta,v)=\frac{1}{2}<v,v>,$$

and thus

$$B(v,v)=\frac{1+h}{2}<v,v>-(1+h)<v,v>_- + ||v||^2 + h||v_\beta||^2$$

$$=\frac{1+h}{2}(<v,v>_+ - <v,v>_-)+||v||^2+h||v_\beta||^2$$

$$=\frac{1+h}{2}|v|^2+||v||^2+h||v_\beta||^2,$$

which proves the desired equality. □

We can now prove an error estimate for the streamline diffusion method (9.28).

Theorem 9.2 There is a constant C such that if u^h satisfies (9.28) and u satisfies (9.13), then

$$(9.30)\qquad ||u-u^h||_\beta\leqslant Ch^{r+1/2}||u||_{r+1}.$$

Proof Let $\tilde{u}^h\in V_h$ be an interpolant of u satisfying (9.16). Writing as before $\eta^h=u-\tilde{u}^h$ and $e^h=u^h-\tilde{u}^h$, and using Lemma 9.2 with v=e and (9.29) with $v=u^h-\tilde{u}^h$, we get

$$||e||_\beta^2=B(e,e)=B(e,\eta^h)-B(e,e^h)=B(e,\eta^h)$$

$$=(e_\beta,\eta^h)+h(e_\beta,\eta_\beta^h)+(e,\eta^h)+h(e,\eta_\beta^h)-(1+h)<e,\eta^h>_-$$

$$\leqslant\frac{h}{4}||e_\beta||^2+h^{-1}||\eta^h||^2+\frac{h}{4}||e_\beta||^2+h||\eta_\beta^h||^2+\frac{1}{4}||e||^2+||\eta^h||^2$$

$$+\frac{1}{4}||e||^2+h^2||\eta_\beta||^2+\frac{(1+h)}{4}|e|^2+(1+h)|\eta^h|^2.$$

Recalling the approximation result (9.16), we thus have

$$\|e\|_\beta^2 \leq Ch^{2r+1}\|u\|_{r+1}^2,$$

which proves the desired estimate. □

Remark 9.2 Notice that it is the presence of the term $h\|e_\beta\|^2$ in the quantity $\|e\|_\beta^2$ dominated by B(e, e) that makes it possible to split the critical term (e_β, η^h) into one term $h\|e_\beta\|^2/4$ that can be "hidden" in $\|e\|_\beta^2$, and one term $h^{-1}\|\eta^h\|^2$ that eventually will produce the factor $h^{r+1/2}$. Thus, the streamline diffusion method has "extra stability" in the streamline direction as compared with the standard Galerkin method where the form b(v,v) dominates only the L_2-norms $\|v\|^2$ and $|v|^2$ and where the quantity $\|\eta_\beta^h\|$ appears in the proof of the error estimate (cf the proof of Theorem 9.1). □

The error estimate (9.30) for the streamline diffusion method (9.28) states that

$$\|u-u^h\| \leq Ch^{r+1/2}\|u\|_{r+1},$$

$$\|(u_\beta-u_\beta^h)\| \leq Ch^r\|u\|_{r+1}.$$

Thus, the L_2-error is half a power of h from being optimal (cf (9.16a)), while the L_2-error of the derivative in the streamline direction is in fact optimal. These estimates indicate that the streamline diffusion method (9.28) should be somewhat better than the standard Galerkin method (9.22) if the exact solution is smooth, but they do not explain the dramatic improvement one actually finds when the exact solution is non-smooth. The fact that the streamline diffusion method also performs well in this latter more difficult case is related to the fact that in this method effects are propagated approximately as in the continuous problem, i e, essentially along the characteristics. One can prove for the streamline diffusion method (see [JNP]) that the effect of a source at a certain point $P \in \Omega$ decays at least as rapidly as $\exp(-d/C\sqrt{h})$ where d is the distance to P in directions perpendicular to the characteristics ("crosswind" directions), and like $\exp(-d/Ch)$ in the direction opposite to the characteristics ("upwind" direction). In particular, this means that the effect of e g a jump in the exact solution across a characteristic will be limited to a narrow region around the characteristic of width at most $0(\sqrt{h})$ (in certain cases the width is improved to $0(h^{3/4})$, see [JSW]). On the other hand, in the standard Galerkin method effects may propagate in the crosswind and even in the upwind direction with little damping (see the discussion for the one-dimensional problem (9.19) in Example 9.1).

Remark 9.3 Note that for the continuous problem (9.13) with g=0 for simplicity, we have the following stability estimate:

$$|u|+\|u\|+\|u_\beta\|\leq C\|f\|.$$

This estimate follows by multiplying (9.13) by u which gives control of $|u|$ and $\|u\|$, and the control of u_β then follows through the equation $u_\beta=f-u$. In the streamline diffusion method (9.28) the corresponding stability estimate, obtained by taking $v=u^h$ and using Lemma 9.2, reads $\|u^h\|_\beta\leq C\|f\|$ or

$$|u^h|+\|u^h\|+\sqrt{h}\ \|u^h_\beta\|\leq C\|f\|.$$

This estimate is a weaker variant of the above estimate for the continuous problem with less control of the streamline derivative. In the discrete case we have no equation analogous to $u_\beta=f-u$ and hence control of $\|u^h_\beta\|$ does not follow from control of $\|u^h\|$. Instead, in the streamline diffusion method, partial control of $\|u^h_\beta\|$ is explicitly built in through the modified test function $v+hv_\beta$. Notice also that in the standard Galerkin method the stability estimate reads $|u^h|+\|u^h\|\leq C\|f\|$, with no extra control of $\|u^h_\beta\|$. Here one can only guarantee $\|u^h_\beta\|\leq Ch^{-1}\|u^h\|\leq Ch^{-1}\|f\|$ through an inverse estimate, cf (7.47). □

9.7.2 The streamline diffusion method with $\varepsilon>0$

Let us start from the stationary problem (9.12) with g=0 and $h>\varepsilon>0$. Multiplying the equation $-\varepsilon\Delta u+u_\beta+u=f$ by the test function $v+\delta v_\beta$, where $v\in H^1_0(\Omega)$, and integrating, we get

$$-\varepsilon\delta(\Delta u,v_\beta)+\varepsilon(\nabla u,\nabla v)+(u_\beta+u,v+\delta v_\beta)=(f,v+\delta v_\beta),$$

where the term $-\varepsilon(\Delta u, v)$ has been integrated by parts. Here δ is a positive parameter to be specified below. To formulate a discrete analogue of this relation by replacing u by $u^h\in\mathring{V}_h$ and restricting v to $\mathring{V}_h$, we have to give a suitable meaning to the term $(\Delta u^h, v_\beta)$, since this expression is not directly well-defined for $u^h, v\in\mathring{V}_h$. The correct definition turns out to be simply the following in this case:

$$(9.31)\qquad (\Delta u^h, v_\beta)\equiv\sum_{K\in T_h}\int_K\Delta u^h v_\beta dx,$$

ie, we just sum the integrals over the interior of each triangle K where Δu^h and v_β are well-defined. We now formulate the following streamline diffusion method for (9.12): Find $u^h\in\mathring{V}_h$ such that

$$\text{(9.32)} \qquad \varepsilon(\nabla u^h,\nabla v)-\varepsilon\delta(\Delta u^h, v_\beta)+(u^h_\beta+u^h,v+\delta v_\beta)=(f,v+\delta v_\beta) \quad \forall v\in \mathring{V}_h,$$

where $\delta=\bar{C}h$ if $\varepsilon<h$ with $\bar{C}>0$ sufficiently small (see Remark 9.4), and $\delta=0$ if $\varepsilon\geqslant h$. Clearly, this is a consistent formulation since (9.32) is satisfied with u_h replaced by u as we noted above. The error estimate (9.30) and the localization results for (9.28) can be extended to the method (9.32) with $\varepsilon<h$. To sum up, the method (9.32) is an answer to the problem of constructing a higher-order accurate method for (9.12) with good stability properties (cf Remark 9.6).

Remark 9.4 Let us give a proof of the basic stability estimate for (9.32) in the case $\varepsilon<h$, which proves in particular that the presence of the term $-\varepsilon\delta(\Delta u^h, v_\beta)$ does not degrade the extra stability introduced by the term $\delta(u^h_\beta, v_\beta)$ if the constant $\bar{C}$ is small enough. By the inverse estimate (7.51) we have for $v\in\mathring{V}_h$

$$|\varepsilon\delta(\Delta v, v_\beta)|\leqslant\frac{1}{2}\varepsilon||\nabla v||^2+\frac{1}{2}\varepsilon\delta C^2h^{-2}\delta||v_\beta||^2,$$

so that with $B_\varepsilon(.\ ,\ .)$ denoting the bilinear form associated with (9.32).

$$B_\varepsilon(v, v)\geqslant\frac{1}{2}\varepsilon||\nabla v||^2+||v||^2+(1-\frac{1}{2}\varepsilon\delta C^2h^{-2})\ \delta||v_\beta||^2, \quad \text{for } v\in\mathring{V}_h.$$

Thus, if $\bar{C}$ is so small that

$$\varepsilon\delta C^2h^{-2}=C^2\bar{C}\varepsilon hh^{-2}\leqslant C^2\bar{C}<1,$$

then

$$B_\varepsilon(v, v)\geqslant\frac{1}{2}(\varepsilon||\nabla v||^2+\delta||v_\beta||^2+||v||^2) \qquad \text{for } v\in\mathring{V}_h,$$

which proves the desired stability result. □

Remark 9.5 If $\{T_h\}$ is not quasi-uniform or $|\beta|$ is variable, then in (9.32) we choose $\delta=\bar{C}h_K/|\beta|$ on $K\in T_h$ if $\varepsilon<h_K|\beta|$, where h_K is the diameter of K, and $\delta=0$ if $\varepsilon\geqslant h_K|\beta|$.

Remark 9.6 As noted above the streamline diffusion method will capture a jump discontinuity of the exact solution in a thin numerical layer. However, within this numerical layer the approximate solution may exhibit over- and under-shoots (cf Fig 9.3). Recently, in [HFM] and [HM2], a modified streamline diffusion method with improved shock-capturing properties (reduced over- and under-shoots) was introduced. In this method the test functions are modified as follows:

$$v+\delta\beta\cdot\nabla v+\bar{\delta}\bar{\beta}\cdot\nabla v,$$

where

$$\bar{\beta}=\frac{\beta\cdot\nabla u^h}{|\nabla u^h|^2}\nabla u^h,$$

i e $\bar{\beta}$ is the projection of β onto ∇u^h. Since $\bar{\beta}$ depends on the unknown discrete solution u^h, this leads to a non-linear method even though the underlying problem is linear. Further, as above $\delta=O(h/|\beta|)$ and also $\bar{\delta}=O(h/|\bar{\beta}|)$. For numerical results see Examples 9.2, 9.6 and 13.9. The problem of theoretically explaining the improved shock-capturing of the modified method is considered in [Sz]. □

Remark 9.7 The streamline diffusion method for (9.12) and (9.13) is basically obtained by multiplication with test functions of the form $v+hv_\beta$ where $v\in V_h$. This means that the test functions belong to a space which is different from the space of *trial functions* V_h where the discrete solution u^h is sought. Such a method, where the test functions are different from the trial functions, is sometimes called a *Petrov-Galerkin* method. Note that in a standard Galerkin method the spaces of trial and test functions are the same (modulo boundary conditions). □

Example 9.2. Consider the convection-diffusion problem

$$\begin{aligned} -\varepsilon\Delta u+\beta\cdot\nabla u&=0 \text{ in } \Omega,\\ u&=g \text{ on } \Gamma_1,\\ \frac{\partial u}{\partial u}&=0 \text{ on } \Gamma_2, \end{aligned}$$

where $\varepsilon=10^{-3}$, $\beta=(\cos 10°, \sin 10°)$, $\Omega=\{x\in\mathbf{R}^2\colon 0<x_i<1\}$ is the unit square, $\Gamma_2=\{x\in\Gamma\colon x_2=1\}$ and $\Gamma_1=\Gamma\setminus\Gamma_2$ where Γ is the boundary of Ω. Further, $g=1$ for $1/2<x_2<1$, $x_1=0$, and $g=0$ if $x_2<1/2$ or $x_1=1$. In Fig 9.7a we give the approximate solution of this problem obtained using the streamline diffusion method on the indicated mesh with $\delta=h$ and using piecewise linear basis functions. In Fig 9.7b we give the corresponding result using a small amount of shock-capturing ($\bar{\delta}=0.15$ h). □

Problems

9.5 Formulate the streamline diffusion method for the one dimensional problem (9.19) with $\varepsilon>0$ as well as $\varepsilon=0$. Determine the corresponding difference schemes in the case of piecewise linear trial functions and

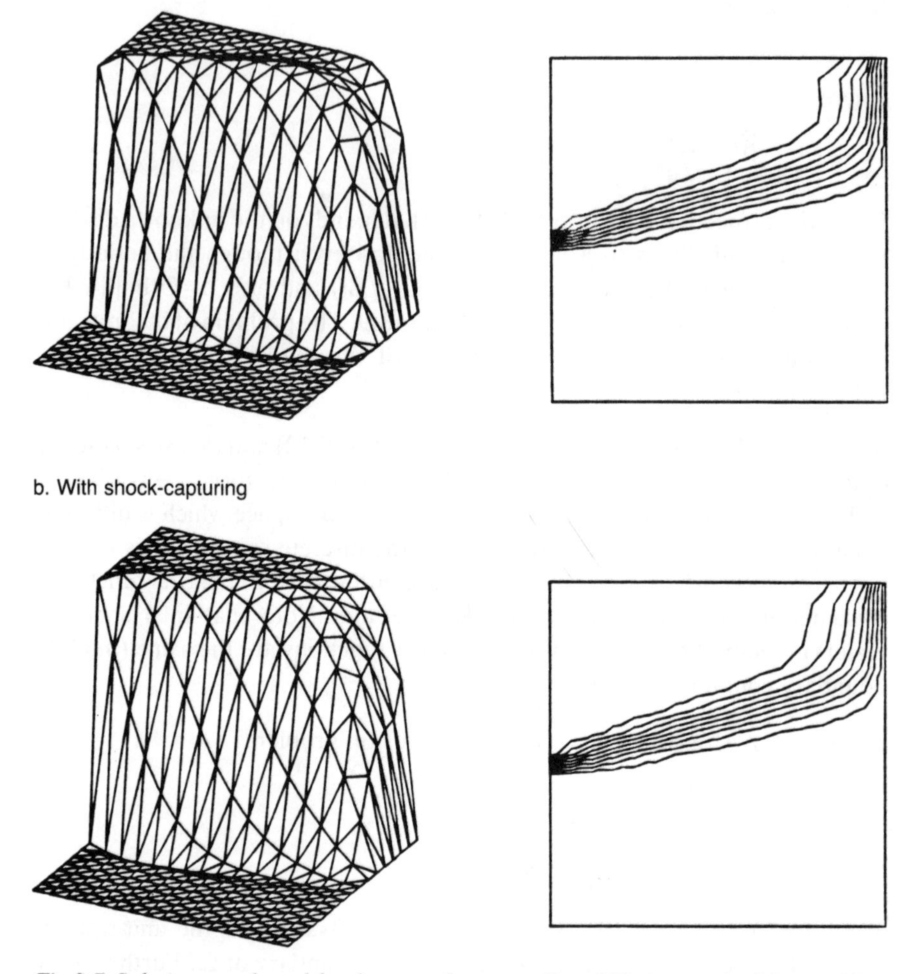

Fig 9.7 Solution graph and level curves for streamline diffusion method for problem in Example 9.2 (a) without and (b) with shock-capturing

a uniform partition. Make a computational comparison with the method (9.20) (cf Fig 9.6 where the thin curve gives the streamline diffusion solution in the case $\varepsilon=0.01$, $h=1/11$ and $\delta=2h/3$). □

9.6 Prove the error estimate (9.30) for the method (9.32) with $\varepsilon<h$. □

9.7 Generalize the streamline diffusion method (9.32) to the variable coefficient problem of Problem 9.4. Hint: Use e g the test function $v+\delta\mathrm{div}(\beta v)$. □

9.8 The discontinuous Galerkin method

Galerkin methods using continuous trial functions will lead to globally coupled systems of linear equations, i e, systems where a change in data at one node will (at least in principle) affect the solution at all nodes. This is natural in the elliptic problem (9.12) with $\varepsilon>0$, but not so for the purely hyperbolic problem (9.13) with $\varepsilon=0$. In this latter case it would be more natural to be able to solve the linear system by successive elimination starting at the inflow boundary Γ_-.

We will now consider a finite element method for the reduced problem (9.13) which permits such a solution procedure and which has stability and convergence properties similar to that of the streamline diffusion method. This method, the *discontinuous Galerkin method,* may be viewed as a generalization of the method with the same name in Chapter 8. It is based on using the following finite element space:

$$W_h=\{v\in L_2(\Omega):\ v|_K\in P_r(K)\quad \forall K\in T_h\},$$

that is, the space of piecewise polyomials of degree $r\geq 0$ with no continuity requirements across interelement boundaries.

To define this method let us first introduce some notation. For $K\in T_h$ we split the boundary ∂K of the triangle K into an inflow part ∂K_- and an outflow part ∂K_+ defined by

$$\partial K_-=\{x\in\partial K:\ n(x)\cdot\beta<0\},$$

$$\partial K_+=\{x\in\partial K:\ n(x)\cdot\beta\geq 0\},$$

where $n(x)$ is the outward unit normal to ∂K at $x\in K$, (cf Fig 9.8).

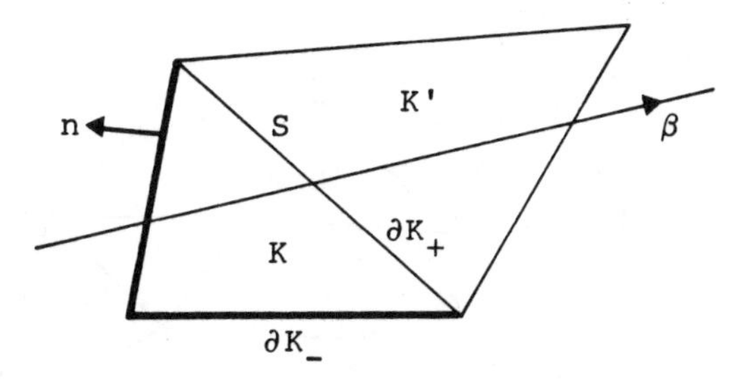

Fig 9.8

Further, suppose S is a side common to two triangles K and K' (cf Fig 9.8) and consider a function $v\in W_h$ which may have a jump discontinuity across S. We define the left and right hand limits v_- and v_+ by

$$v_-(x)=\lim_{s\to 0^-} v(x+s\beta),$$

$$v_+(x)=\lim_{s\to 0^+} v(x+s\beta),$$

for $x\in S$ and we also define the jump $[v]$ across S by

$$[v]=v_+-v_-.$$

The discontinuous Galerkin method for (9.13) can now be formulated as seeking a function $u^h\in W_h$ according to the following rule: For $K\in T_h$, given u^h_- on ∂K_- find $u^h\equiv u^h|_K\in P_r(K)$ such that

$$(9.33)\qquad (u^h_\beta+u^h,v)_K-\int_{\partial K_-} u^h_+ v_+\, n\cdot\beta\, ds=(f,v)_K-\int_{\partial K_-} u^h_- v_+\, n\cdot\beta\, ds,$$

$$\forall v\in P_r(K),$$

where

$$(w,v)_K=\int_K wv\, dx,\quad u^h_-=g \text{ on } \Gamma_-.$$

To see that this problem admits a unique solution, note that (9.33) is nothing but the standard Galerkin method (9.22) with weakly imposed boundary conditions in the case of just *one element*. Thus, if u^h_- is given on ∂K_- we know that $u^h|_K$ is uniquely determined by (9.33). Now, we can start to determine u^h on the triangles K with $\partial K_-\subset\Gamma_-$ since then $u^h_-=g$ is given. This will then define u^h on the triangles K next to Γ_-, and we may continue this process until u^h has been determined in the whole domain (cf Fig 9.9 where the order in which u^h may be calculated is indicated).

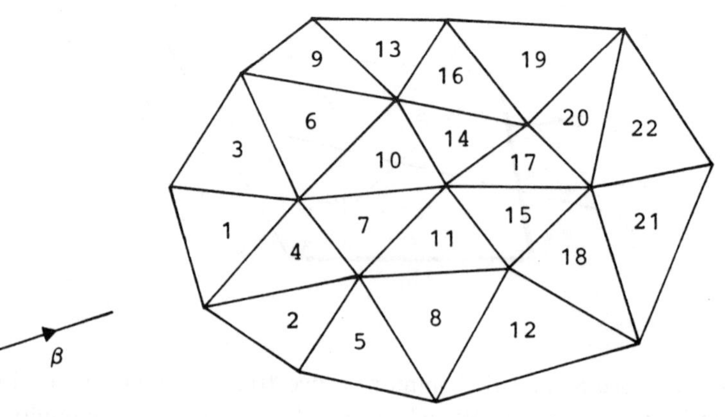

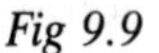

Fig 9.9

To write (9.33) in more compact form suitable for analysis, note that (9.33) can be written

$$B_K(u^h,v)=(f,v)_K \qquad \forall v\in P_r(K),$$

where

$$B_K(w,v)=(w_\beta+w,v)_K-\int_{\partial K_-}[w]v_+n\cdot\beta\,ds.$$

The discontinuous Galerkin method can now be formulated: Find $u^h\in W_h$ such that

$$(9.34)\qquad B(u^h,v)=(f,v) \qquad \forall v\in W_h,$$

where

$$B(w,v)=\sum_{K\in T_h} B_K(w,v),$$

and $u^h_-=g$ on Γ_-. Clearly the exact solution u satisfies the equation $B(u,v)=(f,v)$, $\forall v\in W_h$ (note that $[u]n\cdot\beta=0$), and thus we have the error equation

$$(9.35)\qquad B(u-u^h,v)=0 \qquad \forall v\in W_h.$$

Before analyzing the method (9.34) in some detail let us consider the following examples.

Example 9.3 Let us consider the one-dimensional analogue of (9.13), i e, the problem

$$(9.36)\qquad \begin{aligned} &u_x+u=f \qquad \text{for } 0<x<1,\\ &u(0)=g. \end{aligned}$$

Let $0=x_0<x_1\ .\ .\ .<x_N=1$ be a subdivision of $I=(0,\ 1)$ into subintervals $I_j=(x_j,\ x_{j+1})$. The method (9.33) reads in this case (cf. (8.36)): For $j=0,\ 1,\ .\ .\ .\ N-1$, given $u^h(x_j)_-$ find $u^h=u^h|_{I_j}\in P_r(I_j)$ such that

$$\int_{I_j}(u^h_x+u^h)v\,dx+(u^h(x_j)_+-u^h(x_j)_-)v(x_j)_+=\int_{I_j} fv\,dx \qquad \forall v\in P_r(I_j),$$

where $v(x)_\pm=\lim_{y\to 0\pm} v(x+y)$ and $u^h(x_0)_-=g$. In particular, for $r=0$ in which case u^h is piecewise constant, we get the following method: Find $U_j\equiv u^h(x_j)_-$ such that

$$\frac{U_{j+1}-U_j}{h_j}+U_{j+1}=\frac{1}{h_j}\int_{I_j} f\,dx \qquad j=0,\ .\ .\ .,\ N-1,$$

$$U_0=g,$$

where $h_j=x_{j+1}-x_j$, which is a simple finite difference method for (9.36), namely the *upwind* or *backward Euler* method. □

Example 9.4 Let us consider the method (9.33) in the case $r=0$ and to simplify further, let us assume that $f=0$ and also that the u-term is not present so that we simply have the problem $u_\beta=0$ in Ω, $u=g$ on Γ_-. The discontinuous Galerkin method then reads: For $K\in T_h$, given u^h_- on ∂K_- find the constant $U_K\equiv u^h|_K$ such that

$$-\int_{\partial K_-} U_K\, n\cdot\beta\, ds = -\int_{\partial K_-} u^h_- n\cdot\beta\, ds,$$

ie

$$U_K = \int_{\partial K_-} u^h_- n\cdot\beta\, ds \Big/ \int_{\partial K_-} n\cdot\beta\, ds. \tag{9.37}$$

In other words, for each K the value U_K is obtained as a weighted average of the values of u^h_- on adjoining elements with sides on ∂K_-. As an example, using quadrilateral elements in the following configuration

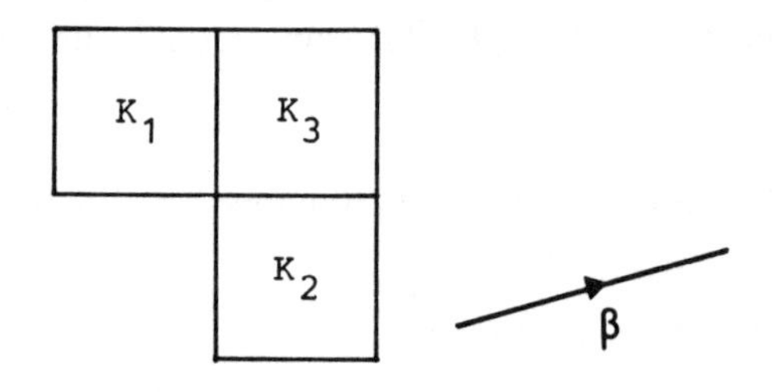

and assuming that $\beta_i>0$, we find that

$$U_3 = \frac{\beta_1}{\beta_1+\beta_2} U_1 + \frac{\beta_2}{\beta_1+\beta_2} U_2,$$

where $U_i=U_{K_i}$. Again this corresponds to a simple difference scheme for the equation $u_\beta=0$ (in fact this is a usual *upwind difference scheme* for this equation if we relate the value U_i to the midpoint of each K_i). □

Let us now prove a stability inequality for the discontinuous Galerkin method (9.34) using the norm $|\cdot|_\beta$ defined by

$$|v|_\beta^2 = \|v\|^2 + \frac{1}{2}\sum_K \int_{\partial K_-} [v]^2 |n\cdot\beta|\, ds + \frac{1}{2}\int_{\Gamma_+} v_-^2 n\cdot\beta\, ds.$$

Lemma 9.3 For any piecewise smooth function v we have

$$B(v, v)=|v|_\beta^2-\frac{1}{2}\int_{\Gamma_-} v_-^2|n\cdot\beta|\, ds.$$

Proof By Green's formula

$$2(v_\beta, v)_K=\int_{\partial K_+} v_-^2 n\cdot\beta\, ds-\int_{\partial K_-} v_+^2|n\cdot\beta|ds,$$

and thus

$$2B(v,v)=\sum_K\{\int_{\partial K_+} v_-^2 n\cdot\beta\, ds-\int_{\partial K_-} v_+^2|n\cdot\beta|ds$$

$$+2\int_{\partial K_-}(v_+-v_-)v_+|n\cdot\beta|ds\}+2||v||^2.$$

Since every side of ∂K_+ coincides with a side of $\partial K'_-$ for an adjoining element K', except if $\partial K_+\subseteq\Gamma_+$, and similarly with + and − reversed, we have

$$\sum_K\int_{\partial K_+} v_-^2 n\cdot\beta\, ds=\sum_K\int_{\partial K_-} v_-^2|n\cdot\beta|ds$$

$$+\int_{\Gamma_+} v_-^2 n\cdot\beta\, ds-\int_{\Gamma_-} v_-^2|n\cdot\beta|ds,$$

and consequently

$$2B(v,v)=\sum_K\{\int_{\partial K_-}(v_+^2-2v_-v_++v_-^2)|n\cdot\beta|ds\}+$$

$$+\int_{\Gamma_+} v_-^2 n\cdot\beta\, ds-\int_{\Gamma_-} v_-^2|n\cdot\beta|ds+2||v||^2,$$

which proves the desired result. □

From Lemma 9.3 we obtain in the usual way existence and uniqueness of a solution to the discontinuous Galerkin scheme (9.34), and it is also possible to derive an error estimate which proves $0(h^r)$ convergence in the $|\cdot|_\beta$-norm. However, this estimate is not the best possible. One can prove that if $\delta=\bar{C}h$ for some suitable constant $\bar{C}$, then for $v\in W_h$

$$B(v,v+\delta v_\beta)\geq C(||v||_\beta^2-\int_{\Gamma_-} v_-^2|n\cdot\beta|ds), \tag{9.38}$$

where

$$||v||_\beta^2=|v|_\beta^2+h\sum_K||v_\beta||_K^2, \qquad ||w||_K^2=(w,w)_K.$$

Using this improved stability, it is possible to prove the following error estimate

(9.39) $\qquad ||u-u^h||_\beta \leqslant Ch^{r+1/2}||u||_{r+1}.$

In the case $r=0$ we have $||v||_\beta=|v|_\beta$ for $v\in W_h$ since here $v_\beta=0$ on each K. Thus, for $r=0$ the stability of Lemma 9.3 should be sufficient to obtain $0(h^{1/2})$ convergence. Let us prove that this is in fact the case.

Theorem 9.3 There is a constant C such that if u^h satisfies (9.34) with $r=0$ and u satisfies (9.13), then

$$|u-u^h|_\beta \leqslant Ch^{\frac{1}{2}}||u||_1.$$

Proof Let $\tilde{u}^h\in W_h$ be the interpolant of u defined by letting $\tilde{u}^h|_K$ be the mean value of u over K for each $K\in T_h$, and let us write as usual $\eta^h=u-\tilde{u}^h$. Applying Lemma 9.3 with $v=e\equiv u-u^h$ and noting that $e_-=0$ on Γ_-, we get, using also the error equation (9.35) with $v=u_h-\tilde{u}_h$

$$\begin{aligned}|e|_\beta^2 &= B(e, e) = B(e, u-\tilde{u}^h)+B(e, \tilde{u}^h-u^h)\\ &= B(e, \eta^h) = \sum_K \{(e_\beta, \eta^h)_K - \int_{\partial K_-} [e]\eta_+^h n\cdot\beta\, ds\} + (e, \eta^h).\end{aligned}$$

Now, $e_\beta=(u-u^h)_\beta=u_\beta$ on each K since u^h is piecewise constant, and thus by Cauchy's inequality

$$\begin{aligned}(9.40) \qquad |e|_\beta^2 \leqslant\; & ||u_\beta||\ ||\eta^h||+||e||\ ||\eta^h||\\ & +(\sum_K \int_{\partial K_-} [e]^2|n\cdot\beta|ds)^{1/2}\cdot(\sum_K \int_{\partial K_-} |\eta^h|^2|n\cdot\beta|ds)^{1/2}.\end{aligned}$$

It is easy to realize that if $u\in C^1(\bar{\Omega})$ so that $\frac{\partial u}{\partial x_i}$, $i=1, 2$, is bounded on $\bar{\Omega}$, then

$$\max_{x\in\bar{\Omega}} |\eta^h(x)|\leqslant Ch,$$

and therefore, since the length of ∂K_- is $0(h)$ and the number of elements is $0(h^{-2})$, we have

$$\sum_K \int_{\partial K_-} |\eta^h|^2|n\cdot\beta|ds \leqslant \sum_K Ch^3 \leqslant Ch.$$

Thus from (9.40) we conclude, hiding terms as usual on the left hand side, that

$$|e|_\beta^2 \leqslant C||u_\beta||h+Ch^2+Ch\leqslant Ch,$$

or

$$|e|_\beta \leqslant Ch^{1/2},$$

where the constant C depends on max $\{|D^\alpha u(x)|: |\alpha|=1, x\in\bar{\Omega}\}$. This proves the desired error estimate, modulo the fact that we have used a somewhat stronger norm on the exact solution u than stated in the theorem. It is in fact easy to see that the norm $\|u\|_1$ is sufficient and this is left to the interested reader. □

Remark 9.8 Suppose we stop the calculation of u^h when u^h has been determined on a subset T_h' of T_h, e g on the triangles $K_1, \ldots, K_{11}$ in Fig 9.10, and let Ω' be the union of the triangles in T_h'. Then clearly the error estimate (9.39) holds with Ω replaced by Ω'. In particular this means that we obtain error estimates in the weighted L_2-norm

$$\int_{\Gamma'_+} v_-^2 n\cdot\beta \, ds,$$

extended over the outflow boundary Γ_+' of each subdomain Ω', e g along the line ABCD of Fig 9.10. □

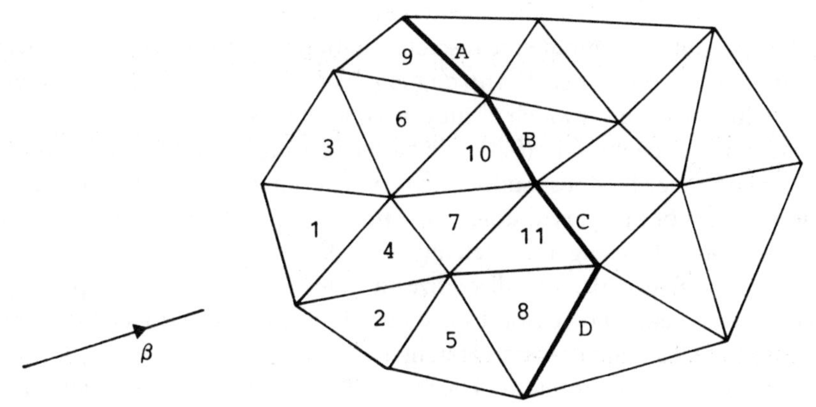

Fig 9.10

We now turn to the time-dependent model problem (9.15) with $\varepsilon=0$, i e, the problem

$$\begin{aligned} &u_t+u_x=f && \text{in } \Omega=J\times I\equiv(0,1)\times(0,T), \\ (9.41) \qquad &u(x,0)=u_0(x) && \text{for } x\in J, \\ &u(0,t)=g(t) && \text{for } t\in I. \end{aligned}$$

As already noted this problem has the same form as the stationary model problem (9.13) with $\varepsilon=0$. Thus, we can apply the discontinuous Galerkin method using a triangulation of $\Omega=J\times I$, ie, a triangulation in *space-time*. It is natural to consider a triangulation where the elements are organized in time as in the following example (note that triangulations in adjoining strips do not necessarily have to match):

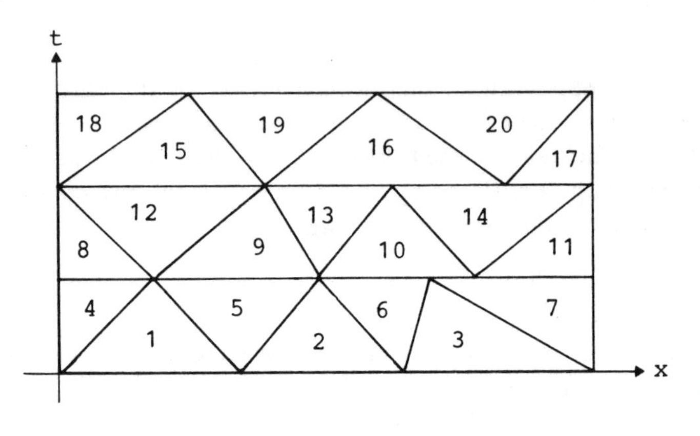

Fig 9.11

It is then possible to compute the discrete solution u^h successively on one strip after another starting for each strip on the left and moving triangle by triangle to the right (the order in which u^h may be computed is indicated in Fig 9.11).

Conventional schemes for (9.41) are based on using separate discretizations in space and time. First a semidiscrete problem (an initial value problem for a linear system of ordinary differential equations) is obtained by discretizing the space variable using finite elements or finite differences, and then a difference method is used to discretize in time. However, for the problem (9.41) there is really no reason to distinguish between x and t and it seems most natural to use space-time elements.

To sum up, the discontinuous Galerkin method has theoretical stability and convergence properties similar to those of the streamline diffusion method. In practice it turns out that when applied to eg (9.41) the discontinuous Galerkin method performs somewhat better than the streamline diffusion method. In fact, already for $r=1$ the discontinuous Galerkin for (9.41) performs remarkably well and we know of no (linear) finite difference method that is better (cf Problems 9.12, 9.15 and Example 9.5 below).

Problems

9.8 Determine the difference scheme corresponding to the discontinuous Galerkin method with $r=1$ for the one-dimensional problem of Example 9.1.

9.9 Evaluate U_K in (9.37) in the case of triangular elements in a general configuration. Distinguish between the cases when ∂K_- consists of one and two sides.

9.10 Let u^h be the solution of the discontinous Galerkin method with r=1 for the problem

$$\frac{\partial u}{\partial x_2}=0 \text{ in } \Omega, \qquad u=g \text{ on} \Gamma_-,$$

on the following triangulation:

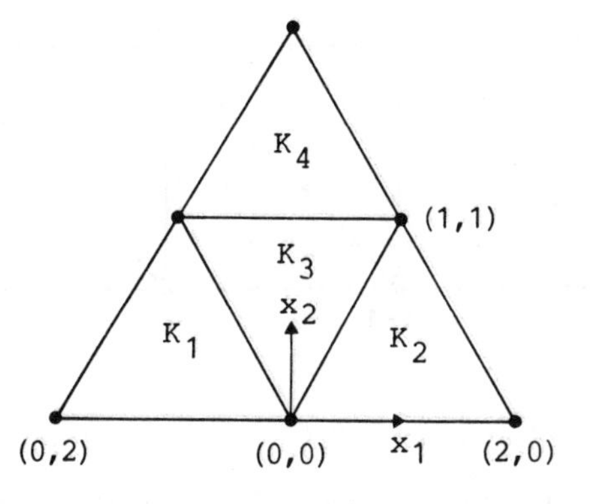

Suppose that u^h_- on $\partial K^1_- \cup \partial K^2_-$ is given by

$$\text{(9.42)} \quad \begin{aligned} u^h_-(x_1, 0)&=\alpha_1+\beta_1 x_1, \qquad & -2<x_1<0,\\ u^h_-(x_1, 0)&=\alpha_2+\beta_2 x_1, \qquad & 0<x_1<2. \end{aligned}$$

Determine u^h_- on ∂K^4_-, i e, assuming that $u^h_-(x_1,1)=\delta_4+\beta_4 x_1$ for $-1<x_1<1$, determine α_4 and β_4 in terms of α_1, β_1, α_2 and β_2. Hint: Prove first that $\frac{\partial u^h}{\partial x_2}=0$ on K^1 and K^2 and then prove that

$$\int_{-1}^{1} u^h_-(x_1,1)v\, dx_1 = \int_{-1}^{1} u^h_-(x_1,0)v\, dx_1 \text{ for } v=1,\ x_1,$$

i e prove that $u^h_-(x_1, 1)$, $-1<x_2<1$, is the L_2-projection onto the space of linear functions on $(-1, 1)$ of the piecewise linear function $u^h_-(x_1, 0)$ given by (9.42) for $-1<x_1<1$.

9.11 Extend the analysis of the previous problem to the more general equation $u_\beta=0$ with β chosen so that the inflow boundary $\Gamma_-=\partial K^1_- \cup \partial K^2_-$. Based on this analysis make an interpretation of the discontinuous Galerkin method for the problem $u_\beta=0$ as a method composed of two steps: *exact transport* and *projection*.

9.12 Consider the discontinuous Galerkin method with r=1 for the problem

$$\frac{\partial u}{\partial t}+\gamma\frac{\partial u}{\partial x}=0, \qquad (x, t)\in R\times R^+,$$

$$u(x, 0)=g(x), \qquad x\in R,$$

with $0\leqslant\gamma\leqslant 1$. Use a uniform triangulation of $R\times R^+$ of the form:

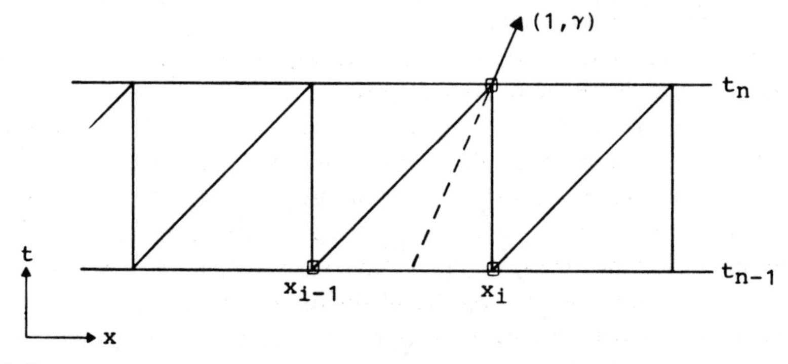

Fig 9.12

where $x_i=ih$, $t_n=nh$, $i=0, \pm1, \pm2, \ldots, n=0, \pm1, \pm2, \ldots$. Suppose we represent the discrete solution $u_-^h(\cdot, t_n)$ on the interval (x_{i-1}, x_i) as

$$u_-^h(x, t_n)=U_i^n+\frac{1}{h}(x-x_i')\,V_i^n, \quad x\in(x_{i-1}, x_i),$$

where $x_i'=x_i-\frac{h}{2}$. Prove, using e g the result of Problem 9.11, that the discontinuous Galerkin method in this case is equivalent to the following explicit difference scheme:

$$\begin{bmatrix} U_i^n \\ V_i^n \end{bmatrix}=(1-\gamma)\begin{bmatrix} 1 & -\frac{1}{2}\gamma \\ 6\gamma & 1-2\gamma-2\gamma^2 \end{bmatrix}\begin{bmatrix} U_i^{n-1} \\ V_i^{n-1} \end{bmatrix}$$
$$+\gamma\begin{bmatrix} 1 & \frac{1}{2}(1-\gamma) \\ -6(1-\gamma) & -3+6\gamma-2\gamma^2 \end{bmatrix}\begin{bmatrix} U_{i-1}^{n-1} \\ V_{i-1}^{n-1} \end{bmatrix},$$

connecting the values of the vector (U, V) associated with the □-marked points in Fig 9.12.

9.13 Prove (9.38) and (9.39).

9.9 The streamline diffusion method for time-dependent convection-diffusion problems

Let us now consider the time-dependent model problem (9.14) with $\varepsilon>0$, i e the problem

$$(9.43)\qquad \begin{aligned} &u_t+u_x-\varepsilon u_{xx}=f && \text{in } \Omega=J\times I,\\ &u(x,0)=u_0(x) && x\in J,\\ &u(x,t)=0 && x=0,1,\ t\in I, \end{aligned}$$

where for simplicity we consider the case of zero boundary data. With $\varepsilon>0$ we cannot apply the discontinuous Galerkin method of the previous section to this problem; to handle the diffusion term $-\varepsilon u_{xx}$, the trial functions should be continuous in the space variable. On the other hand, to be able to compute the discrete solution successively on one time level after the other, it is natural, if we insist on using space-time elements, to use trial functions which are discontinuous in time. Thus, we are led to consider a method where the trial functions are continuous in space and discontinuous in time based on a triangulation of space-time with the elements organized in strips in time e g as in Fig 9.11 or 9.13.

To define such a method let $0=t_0<t_1<\ldots<t_N=T$ be a subdivision of the time interval $I=(0,T)$ and introduce the strips S_n defined by

$$S_n=\{(x,t)\colon 0<x<1,\ t_{n-1}<t<t_n\},$$

for $n=1,\ldots,N$. Further, for each n let V^n be a finite element subspace of $H^1(S_n)$, based on a triangulation of the strip S_n with elements of size $h>\varepsilon$, and let $\mathring{V}^n=\{v\in V^n\colon v(x,t)=0 \text{ for } x=0,1\}$. (Notice that it is not necessary that triangulations of different strips "fit" across the discrete time levels, cf Fig 9.13).

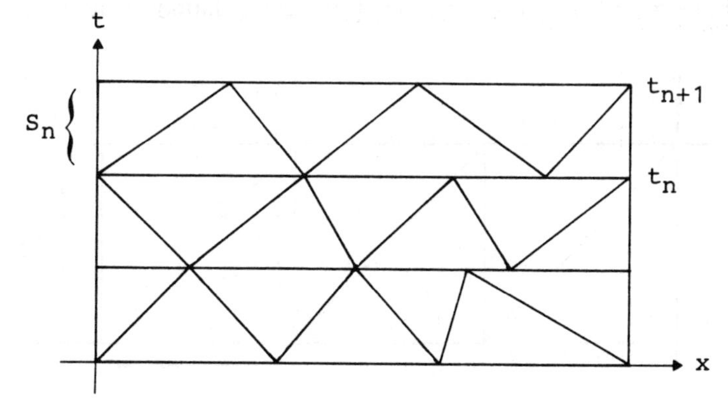

Fig 9.13

If we now apply the streamline diffusion method (9.32) successively on each strip S_n to the problem (9.43), imposing the initial value at $t=t_{n-1}$ weakly and the boundary conditions strongly, we obtain the following method: For $n=1, \ldots, N$, find $u^n \in \mathring{V}^n$ such that

$$(9.44)\qquad \begin{aligned} &(u_t^n+u_x^n,\ v+\delta(v_t+v_x))^n+<u_+^n,\ v_+>^{n-1} \\ &+\varepsilon(u_x^n,\ v_x)^n-\varepsilon\delta(u_{xx}^n,\ v_t+v_x)^n \\ &=(f,\ v+\delta(v_t+v_x))^n+<u_-^{n-1},\ v_+>^{n-1} \qquad \forall v \in \mathring{V}^n, \end{aligned}$$

where $\delta=\bar{C}h$ with $\bar{C}$ sufficiently small for $\varepsilon<h$, $\delta=0$ for $\varepsilon \geqslant h$,

$$(w, v)^n=\int_{S_n} wv\ dxdt, \qquad <w, v>^n=\int_0^1 w(x, t_n)v(x, t_n)dx,$$

$$v_+(x, t)=\lim_{s\to 0^+} v(x, t+s), \qquad v_-(x, t)=\lim_{s\to 0^-} v(x, t+s),$$

and $u_-^0=u_0=$initial data, and $\varepsilon\delta(u_{xx}^n, v_t+v_x)^n$ is defined in a way analogous to (9.31).

For each n (9.44) is equivalent to a linear system of equations and thus we have an *implicit* scheme (cf Problem 9.14). Further, since the space $\mathring{V}^n$ is independently defined on each strip with no continuity requirements from one strip to the other, the solution u^n will in general have jumps across the discrete time levels t_n. In the case $\varepsilon \geqslant h$ and a suitably chosen triangulation of $(0, 1)\times(0, T)$ the method (9.44) would coincide with the discontinuous Galerkin method for parabolic problems presented in Section 8.4.

Problems

9.14 Consider the streamline diffusion method (9.44) with $\varepsilon=0$ and $J=(-\infty, \infty)$ on a regular space-time triangulation of the type

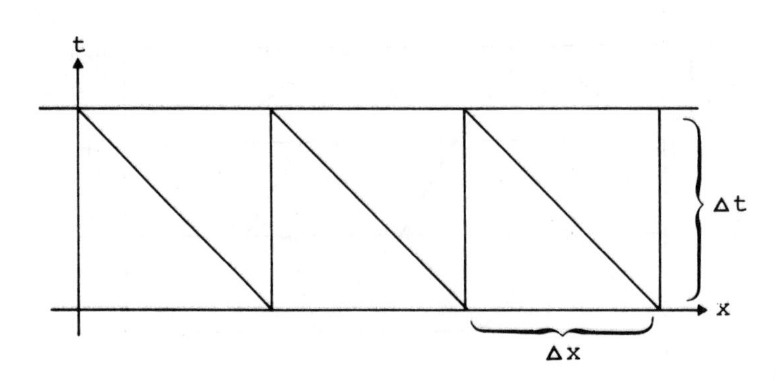

Show that the method in this case is equivalent to the difference scheme:

$$\left(\frac{1}{4}+\frac{2}{3}\lambda+\frac{4}{9}\lambda^2\right)U_{j-2}^{n+1}+\left(\frac{1}{6}\lambda^{-1}-\frac{23}{18}-\frac{101}{36}\lambda-\frac{49}{36}\lambda^2\right)U_{j-1}^{n+1}$$

$$+\left(\frac{2}{3}\lambda^{-1}+\frac{9}{4}+\frac{63}{18}\lambda+\frac{3}{2}\lambda^2\right)U_j^{n+1}+\left(\frac{1}{6}\lambda^{-1}-\frac{5}{6}-\frac{57}{36}\lambda-\frac{25}{36}\lambda^2\right)U_{j+1}^{n+1}$$

$$+\left(\frac{1}{9}+\frac{2}{9}\lambda+\frac{1}{9}\lambda^2\right)U_{j+2}^{n+1}=\left(\frac{1}{6}\lambda^{-1}+\frac{2}{9}+\frac{1}{36}\lambda\right)U_{j-1}^{n}$$

$$+\left(\frac{2}{3}\lambda^{-1}+\frac{3}{4}+\frac{1}{12}\lambda\right)U_j^n+\left(\frac{1}{6}\lambda^{-1}-\frac{1}{3}-\frac{1}{12}\lambda\right)U_{j+1}^n-\left(\frac{5}{36}+\frac{1}{36}\lambda\right)U_{j+2}^n,$$

where $\lambda=\triangle t/\triangle x$ and $U_j^{n+1}=u^n(jh, (n+1)k)_-$.

9.15 Compare computationally the discontinuous Galerkin and the streamline diffusion method for the problem (9.41) with the uniform triangulation of Problem 9.14. Consider the following cases for example:

(i) u is smooth,
(ii) u is piecewise smooth with a jump discontinuity,
(iii) u_0 is a delta function, g=0.

In the same cases also make a computational comparison with the following difference methods:

$$U_j^{n+1}=(1-\lambda)U_j^n+\lambda U_{j-1}^n, \qquad \text{(upwind method)}$$

$$U_j^{n+1}=\left(\frac{\lambda}{2}+\frac{\lambda^2}{2}\right)U_{j-1}^n+(1-\lambda^2)U_j^n-\left(\frac{\lambda}{2}-\frac{\lambda^2}{2}\right)U_{j+1}^n \qquad \text{(Lax-Wendroff)}$$

$$(1-\lambda)U_{j-1}^{n+1}+(1+\lambda)U_j^{n+1}=(1+\lambda)U_{j-1}^n+(1-\lambda)U_j^n \qquad \text{(box scheme)}$$

$$\left.\begin{aligned}V_j=&\frac{\lambda^3}{2}U_{j-2}^n+\left(\frac{1}{6}+\frac{\lambda}{2}+\frac{\lambda^2}{2}-\frac{\lambda^3}{3}\right)U_{j-1}^n\\&+\left(\frac{2}{3}-\lambda^2+\frac{\lambda^3}{2}\right)U_j^n+\left(\frac{1}{6}-\frac{\lambda}{2}+\frac{\lambda^2}{2}-\frac{\lambda^3}{6}\right)U_{j+1}^n\\U_j^{n+1}=&V_j-\frac{1}{6}(V_{j+1}-2V_j+V_{j-1})\end{aligned}\right\}\ \text{(Shasta)}$$

$$U_j^{n+1}=U_j^{n-1}-\lambda U_{j+1}^n+\lambda U_{j-1}^n \qquad \text{(leap-frog)}$$

where U_j^n approximates $u(j\triangle x, n\triangle t)$ and $\lambda=\dfrac{\triangle t}{\triangle x}$.

Example 9.5 In Fig 9.14 below we give the results after 49 time steps obtained by applying the discontinuous Galerkin method of Problem 9.12 and the difference methods of Problem 9.15 to (9.41) with $\lambda=0.56$ and a step function as initial data.

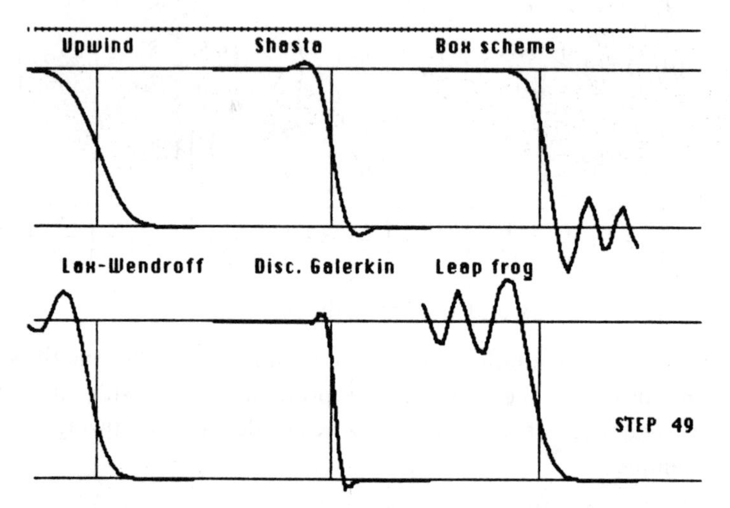

Fig 9.14 Comparison of the discontinuous Galerkin method and some difference schemes for a convection problem

Example 9.6 Consider the convection problem in two space dimensions

$$\frac{\partial u}{\partial t}+\beta\cdot\nabla u=0 \qquad \text{for } x\in\Omega,\ t>0,$$

$$u(x,0)=u_0(x) \qquad \text{for } x\in\Omega,$$

where $\Omega=(-1,1)\times(-1,1)$ and β is the velocity field

$$\beta(x)=r(-\sin\theta,\ \cos\theta), \qquad x=r(\cos\theta,\ \sin\theta),$$

corresponding to a clockwise rotation around the origin. In Fig 9.15a, b we give the initial condition u_0 and the corresponding approximate solution $u^h(\cdot,t)$ after rotation 360° obtained by the direct extension of the streamline diffusion method of Section 9.9. In Fig 9.16a, b we give the initial condition u_0 and the corresponding approximate solution after rotation 90° obtained by the streamline diffusion method with a certain amount of shock-capturing. Note that in this experiment the space-time finite element mesh was adaptively modified automatically in each time step using the technique indicated in Section 4.6. The time step was chosen to be qual to the minimal space step on each computational 'slab' $S_n=\Omega\times I_n$, $I_n=(t_{n-1},\ t_n)$. □

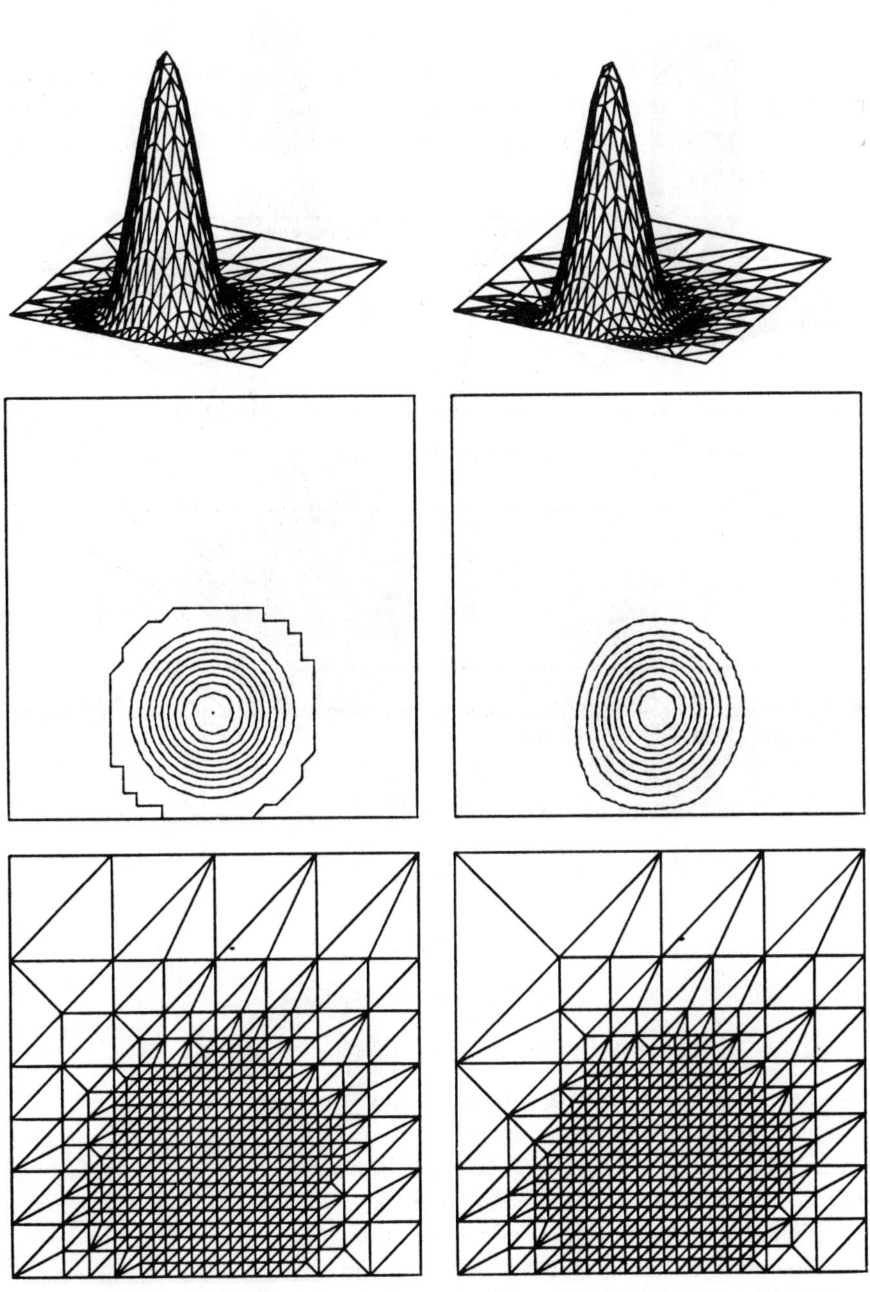

a. Graph and level curves for initial condition together with initial mesh.

b. Same as in a. for finite element solution after rotation 360°.

Fig 9.15 Streamline diffusion method without shock-capturing for convection problem with fairly smooth initial condition

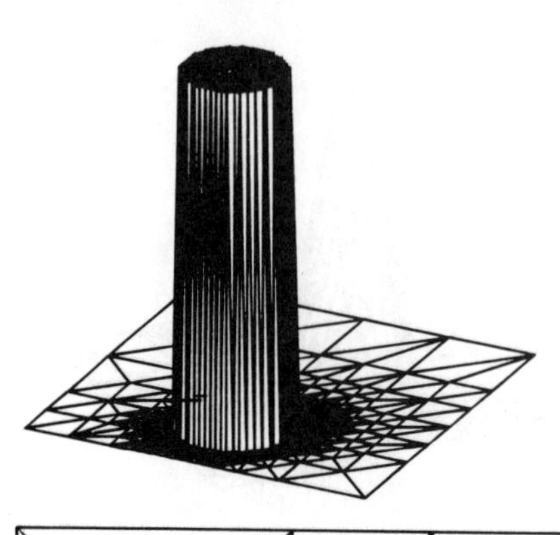

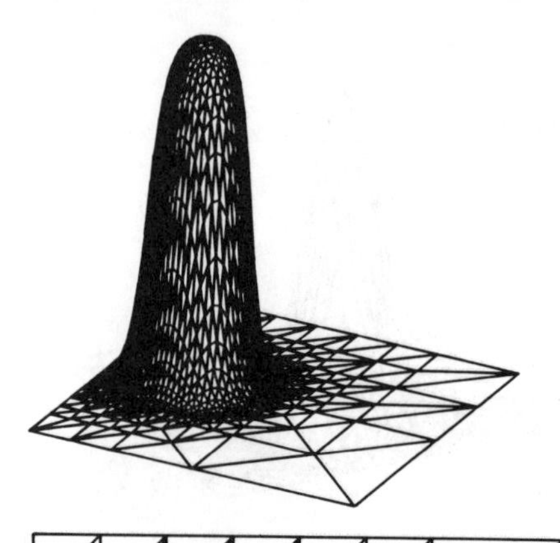

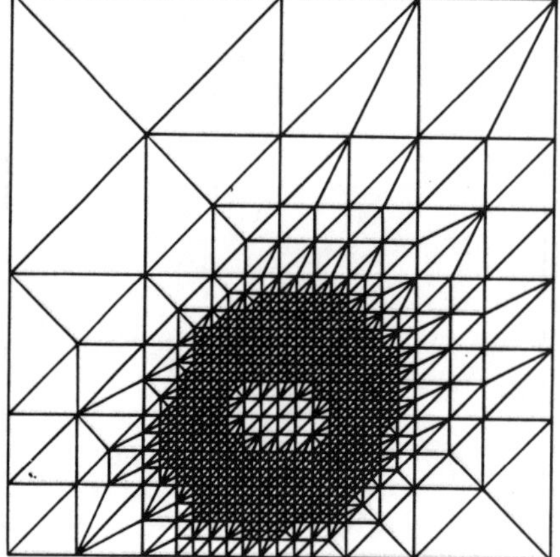

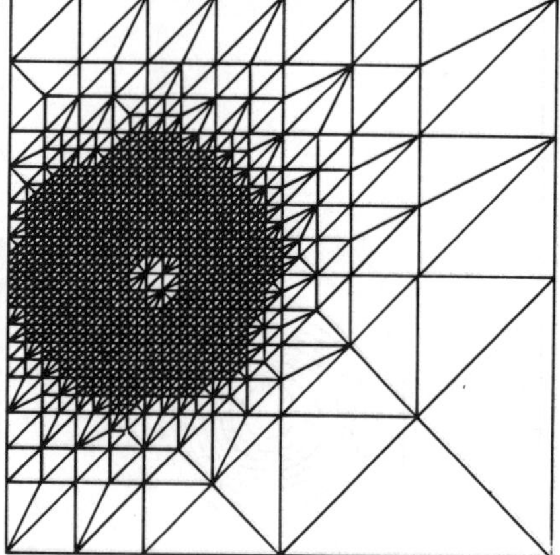

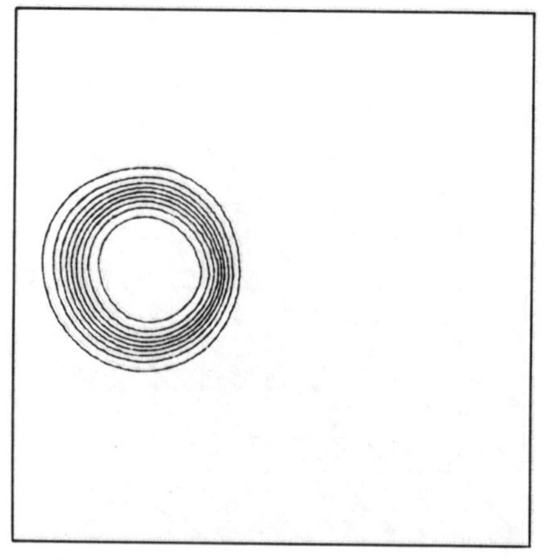

a. Graph of initial condition and initial mesh.

b. Graph and level curves of finite element solution together with mesh after rotation 90°.

Fig 9.16 Streamline diffusion method with shock-capturing for convection problem with non-smooth initial conition

9.10 Friedrichs' systems

9.10.1 The continuous problem

In this section we will briefly indicate how to extend the streamline diffusion and the discontinuous Galerkin methods to the case of linear first order hyperbolic systems of the form (9.8), or Friedrichs' systems (cf [Le]). We then consider first the following problem in a domain $\Omega \subset R^d$ with boundary Γ:

(9.45a) $$Lu \equiv \sum_{i=1}^{d} A_i \frac{\partial u}{\partial x_i} + Ku = F \text{ in } \Omega,$$

(9.45b) $$(M-D)u=0 \qquad \text{on } \Gamma,$$

Here the A_i, K and M are given $m \times m$ matrices depending on x, u is an m-vector and

$$D = \sum_{i=1}^{d} n_i A_i,$$

where $(n_1, \ldots, n_d)$ is the outward unit normal to Γ. We assume that the matrices A_i are symmetric (with real elements) and that

(9.46a) $$M+M^* \geqslant 0 \qquad \text{on } \Gamma,$$

(9.46b) $$K+K^* - \sum_{i=1}^{d} \frac{\partial A_i}{\partial x_i} \geqslant \bar{\sigma} I \qquad \text{in } \Omega,$$

(9.46c) $$\text{Ker } (D-M) + \text{Ker } (D+M) = R^m \qquad \text{on } \Gamma,$$

where $\bar{\sigma}$ is a nonnegative constant, E^* denotes the transpose of a matrix E and I is the identity matrix. For matrices E and F we have written $E \geqslant F$ to mean that E–F is positive semi-definite, and Ker $E = \{\xi \in R^m: E\xi = 0\}$. Under the conditions (9.46) (with $\bar{\sigma} > 0$) and some smoothness assumptions one can prove that if $F \in [L_2(\Omega)]^m$, then (9.45) admits a unique solution, (see [F]).

Many problems in mechanics and physics can be written in the form (9.45). Let us here mention only two special cases.

Example 9.7 The reduced problem (9.13) has the form (9.45) with $m=1$, $A_i = \beta_i$, $K=1$, $D = \beta \cdot n$ and $M = |D|$. □

Example 9.8 The initial-boundary value problem for the wave equation

(9.47a) $$\frac{\partial^2 w}{\partial x_1^2} - \frac{\partial^2 w}{\partial x_2^2} = f \qquad 0 < x_1 < 1, \ |x_2| < 1,$$

(9.47b) $\quad w(x_1, -1)=w(x_1, 1)=0 \qquad 0<x_1<1,$

(9.47c) $\quad w(0, x_2)=\frac{\partial w}{\partial x_1}(0, x_2)=0 \qquad |x_2|<1,$

where x_1 is a time variable, can be written in the form (9.45) with

$\Omega=(0,1)\times(-1, 1)$, $u=(u_1, u_2)$, $u_1=\frac{\partial w}{\partial x_1}$, $u_2=\frac{\partial w}{\partial x_2}$, $F=(f, 0)$ and

$$A_1=\begin{bmatrix} 1 & 0 \\ 0 & 1 \end{bmatrix}, \; A_2=\begin{bmatrix} 0 & -1 \\ -1 & 0 \end{bmatrix}, \; K=\begin{bmatrix} 0 & 0 \\ 0 & 0 \end{bmatrix},$$

$$M=\begin{bmatrix} 1 & 0 \\ 0 & 1 \end{bmatrix} \quad \text{for } x_1=0 \text{ or } x_1=1,$$

$$M=\begin{bmatrix} 2 & 1 \\ -1 & 0 \end{bmatrix} \quad \text{for } x_2=-1, \; M=\begin{bmatrix} 2 & -1 \\ 1 & 0 \end{bmatrix} \quad \text{for } x_2=1.$$

Note that the boundary conditions (9.47b) translate into the conditions $u_1=\frac{\partial w}{\partial x_1}=0$ for $|x_2|=1$, $0<x_1<1$, which correspond to the conditions

$$(M-D)u=\begin{bmatrix} 2 & 0 \\ \pm 2 & 0 \end{bmatrix} \begin{bmatrix} u_1 \\ u_2 \end{bmatrix}=0 \quad \text{for } x_2=\pm 1, \; 0<x_1<1. \quad \square$$

Let us now generalize the standard Galerkin method, the streamline diffusion method and the discontinuous Galerkin method to the Friedrichs' system (9.45). We will use the following notation

$$(v, w)=\int_\Omega v \cdot w \, dx, \quad \|v\|=(v, v)^{1/2},$$

$$\langle v, w\rangle=\int_\Gamma v \cdot w \, ds, \quad |v|=\langle v, v\rangle^{1/2}.$$

By Green's formula we have

$$(Lv, w)=\langle Dv, w\rangle+(v, L^*w),$$

where

$$L^*=-\sum_{i=1}^{d} A_i \frac{\partial}{\partial x_i}-\sum_{i=1}^{d} \frac{\partial A_i}{\partial x_i}+K^*,$$

so that in particular,

(9.48) $$(Lv, v)=\frac{1}{2}((L+L^*)v, v)+\frac{1}{2}\langle Dv, v\rangle,$$

where by (9.46b),

$$L+L^* \geqslant \bar{\sigma} I.$$

We also introduce the spaces $\hat{V}_h=[V_h]^m$ and $\hat{W}_h=[W_h]^m$ with V_h and W_h defined in Sections 9.4 and 9.8 above. We can now formulate our methods for (9.45). For the error estimates we also assume that $M+M^* \geqslant cI$ or Γ for some $c>0$ (cf [JNP]).

9.10.2 The standard Galerkin method

Find $u^h \in \hat{V}_h$ such that

$$(Lu^h, v)+\frac{1}{2} <(M-D)u^h, v>=(F, v) \qquad \forall v \in \hat{V}_h.$$

Choosing here $v=u^h$ and using Green's formula (9.48) we obtain (with $\bar{\sigma}>0$) the stability estimate $||u^h||+|u^h| \leqslant C||F||$, from which error estimates of the form $||u-u^h|| \leqslant Ch^r||u||_{r+1}$ can be derived in the usual way.

9.10.3 The streamline diffusion method

Find $u^h \in \hat{V}_h$ such that

$$(9.49) \qquad (Lu^h, v+hL_0v)+\frac{1}{2} <(M-D)u^h, v>=(F, v+hL_0v) \qquad \forall v \in \hat{V}_h,$$

where

$$L_0= \sum_{i=1}^{d} A_i \frac{\partial}{\partial x_i}.$$

Again choosing $v=u^h$, we obtain (for h sufficiently small) the stability estimate (with again $\bar{\delta}>0$)

$$\sqrt{h}||L_0u^h||+||u^h||+<Mu^h, u^h>^{1/2} \leqslant C||F||,$$

from which we obtain the error estimate $||u-u_h|| \leqslant Ch^{r+1/2}||u||_{r+1}$, arguing as in the proof of Theorem 9.2.

9.10.4 The discontinuous Galerkin method

To formulate this method we need additional notation. For $K \in T_h$ we write

$$(v, w)_K=\int_K v \cdot w \, dx, \quad <v, w>=\int_{\partial K} vw \, ds,$$

$$[w]=w^{int}-w^{ext},$$

$$w^{int}(x)=\lim_{\substack{y\to x\\ y\in K}} w(y),\quad w^{ext}=\lim_{\substack{y\to x\\ y\notin K}} w(y),\quad x\in\partial K,$$

$$D_K=\sum_{i=1}^{d} A_i n_i^K,$$

where $n^K=(n_i^K)$ is the outward unit normal to K, and where we set $w^{ext}(x)=0$ for $x\in\Gamma$. Further, for each $K\in T_h$, we introduce matrices M_K defined on ∂K and satisfying, for $K, K'\in T_h$,

$$M_K=M_{K'} \qquad \text{on } \partial K\cap\partial K',$$

$$M_K+M_K^*\geq 0 \qquad \text{on } \partial K,\ M_K=M \text{ on } \Gamma.$$

Here a possible choice on interior edges is $M_K=\lambda I$ with $\lambda>0$.

The discontinuous Galerkin method for (9.45) can now be formulated: Find $u^h\in\hat{W}_h$ such that

$$\text{(9.50)}\qquad \sum_K \{(Lu^h, v)_K+\frac{1}{2}<(M_K-D_K)[u^h], v>_K\}=(F, v), \qquad \forall v\in\hat{W}_h.$$

For this method we again have error estimates of the form

$$\|u-u^h\|\leq Ch^{r+\frac{1}{2}}.$$

Note that in the scalar case with $m=1$, choosing $M_K=|\beta\cdot n^K|$ in (9.50), where $\beta=(\beta_i)$, $\beta_i=A_i$, gives the discontinuous Galerkin method (9.34).

The formulation (9.45) also includes time-dependent problems if we consider x_1 to be a time variable and choose e g $A_1=I$, $\Omega=(0, T)\times\Omega'$ and $M=I$ for $x_1=0, T$. In this case we should modify the streamline diffusion method (9.49) following the pattern of Section 9.9. On the other hand, the discontinuous Galerkin (9.50) directly applies also to the time-dependent case, cf the following example:

Example 9.9 Let us again consider the one-dimensional wave equation (9.47) written on system form according to Example 9.8, with now $f=0$ and non-zero initial conditions for $x_1=0$:

$$\text{(9.51a)}\qquad \frac{\partial u}{\partial x_1}+A\frac{\partial u}{\partial x_2}=0 \qquad |x_2|<1,\ 0<x_1<1,$$

$$\text{(9.51b)}\qquad (M-D)u=0 \qquad x_2=\pm 1,\ 0<x_1<1,$$

$$\text{(9.51c)}\qquad u(0, x_2)=u^0(0, x_2) \qquad |x_2|<1,$$

where

$$A=\begin{bmatrix} 0 & -1 \\ -1 & 0 \end{bmatrix}, M=\begin{bmatrix} 2 & \mp 1 \\ \pm 1 & 0 \end{bmatrix} \text{ for } x_2=\pm 1.$$

Changing dependent variables through the orthogonal transformation

$$\varphi = Su, \ S=\frac{1}{\sqrt{2}}\begin{bmatrix} 1 & -1 \\ 1 & 1 \end{bmatrix},$$

$$u=S^*\varphi,$$

we can write (9.51) as

$$\text{(9.52a)} \qquad \frac{\partial \varphi}{\partial x_1}+\hat{A}\frac{\partial \varphi}{\partial x_2}=0,$$

$$\text{(9.52b)} \qquad (\hat{M}-\hat{D})\varphi=0 \qquad x_2=\pm 1,\ 0<x_1<1,$$

$$\text{(9.52c)} \qquad \varphi(0, x_2)=\varphi^0(0, x_2) \qquad |x_2|<1,$$

where

$$\hat{A}=SAS^*=\begin{bmatrix} 1 & 0 \\ 0 & -1 \end{bmatrix},$$

$$\hat{M}-\hat{D}\equiv SMS^*-SDS^*=\begin{bmatrix} 2 & 2 \\ 0 & 0 \end{bmatrix} \quad \text{for } x_2=-1,$$

$$\hat{M}-\hat{D}=\begin{bmatrix} 0 & 0 \\ 2 & 2 \end{bmatrix} \quad \text{for } x_2=1.$$

Note that (9.52a) is an *uncoupled* system of two scalar advection equations, and that the coupling in (9.52) only occurs through the boundary conditions.

Suppose we now apply the discontinuous Galekin method to (9.52) with $M_K=\hat{M}_K$ given, on interior edges, by

$$\text{(9.53)} \qquad \hat{M}_K=\begin{bmatrix} |n_1^K+n_2^K| & 0 \\ 0 & |n_1^K-n_2^K| \end{bmatrix} \equiv \begin{bmatrix} m_1^K & 0 \\ 0 & m_2^K \end{bmatrix}.$$

We then obtain a discrete analogue of (9.52) consisting of two discontinuous Galerkin discretizations of scalar advection problems which are coupled only through the boundary conditions (9.52b). Alternatively we may consider a direct application of the discontinuous Galerkin method to the coupled problem (9.51) with the M_K given by

$$M_K=S^*\hat{M}_K S=\begin{bmatrix} m_2^K+m_1^K & m_2^K-m_1^K \\ m_2^K-m_1^K & m_2^K+m_1^K \end{bmatrix},$$

with the notation of (9.53). In fact these two approaches are equivalent and would produce the same numerical results since if $v=S^*\psi$, we have

$$(Au, v)_K=(AS^*\varphi, S^*\psi)_K=(SAS^*\varphi, \psi)_K=(\hat{A}\varphi, \psi)_K,$$

and corresponding relations hold for the remaining terms. Thus, we conclude that the discontinuous Galerkin method applied to the coupled problem (9.51) with proper choice of the M_K "automatically diagonalizes" the system (9.51a). The same holds for the streamline diffusion method. This is of interest e g when analyzing the nature of propagation of effects in the discrete analogues of the coupled problem (9.51). □

Remark 9.9 The streamline diffusion method for stationary problems was introduced in [HB1], [HB2]. The mathematical analysis of the method was started in [JN] and was continued, with extensions to time-dependent problems, in [J2], [Na], [JNP]. The method has also recently been extended with good results to incompressible and compressible flow problems, see [BH], [HFM], [HMM], [HM1], [HM2], [JSa], [J4], [JSz1], [JSz2], [JSzH], [Sz], and Chapter 13. For combinations of finite element methods and methods of characteristics, see [DR], [M], [BPHL]. The discontinuous Galerkin method was first analyzed in [LRa], see also [JP2]. □

Problems

9.16 Apply the discontinuous Galerkin method to (9.51) or (9.52) to compute approximate solutions of the wave equation (9.47). Test the performance of the method with different degrees of regularity of initial data as e g in Problem 9.15. Also compare with the results of other methods for the wave equation such as e g those presented in Section 9.11 below.

9.17 Prove error estimates for the standard Galerkin and the streamline diffusion methods for (9.45) with $\bar{\sigma}>0$ and $M+M^*\geqslant cI$, $c>0$.

9.11 Second order hyperbolic problems

Previously we have considered first order hyperbolic problems. We now turn to second order problems. A typical example is the wave equation:

(9.54a)	$\ddot{u}-\Delta u=f$	in $\Omega\times I$,
(9.54b)	$u=0$	on $\Gamma\times I$,
(9.54c)	$u(x, 0)=u_0(x)$, $\dot{u}(x, 0)=u_1(x)$,	for $x\in\Omega$,

where Ω is a bounded domain in R^2 with boundary Γ, $I=(0,T)$, $\ddot{u}=\dfrac{\partial^2 u}{\partial t^2}$ and $\dot{u}=\dfrac{\partial u}{\partial t}$. This equation models e g a vibrating membrane with given deflection u_0 and initial velocity u_1. Following Example 9.8 it is possible to rewrite (9.54) as a first order hyperbolic system and in principle we may then apply the methods of the previous section. However, with this approach we introduce new unknowns which result in an increase in the number of variables in the discrete problems. Thus, there are good reasons to try to keep the formulation (9.54) involving second derivatives and a scalar unknown. However, with this formulation it does not seem to be known how to construct methods combining good stability with high accuracy. In particular, we cannot extend the streamline diffusion and discontinuous Galerkin methods to the wave equation (9.54) because of the presence of second order differential operators. Anyway let us here describe some methods for (9.54) that are currently used.

The wave equation (9.54) can be given the following variational formulation using the notation of Section 8.2: Find $u(t)\in V=H_0^1(\Omega)$, such that for $t\in I$,

$$(9.55a)\qquad (\ddot{u}(t), v)+ a(u(t), v)=(f(t), v) \qquad \forall v\in V,$$

$$(9.55b)\qquad u(0)=u_0,\ \dot{u}(0)=u_1.$$

The basic energy estimate for (9.55) is obtained by choosing $v=\dot{u}$ in (9.55a) which gives with $f=0$ for simplicity,

$$\frac{1}{2}\,\frac{d}{dt}\|\dot{u}(t)\|^2+\frac{1}{2}\,\frac{d}{dt}\|\nabla u(t)\|^2=0, \qquad t\in I,$$

so that

$$(9.56)\qquad \|\dot{u}(t)\|^2+\|\nabla u(t)\|^2=\text{constant}=\|u_1\|^2+\|\nabla u_0\|^2, \qquad t\in I.$$

This equation expresses the fact that the *total energy* of the system (9.54) is conserved if the applied force $f=0$.

For the numerical solution of (9.55) suppose the finite element space $V_h\subset H_0^1(\Omega)$ is given, and let us first formulate the following *semi discrete analogue:* Find $u^h(t)\in V$ such that for $t\in(0, T)$

$$(9.57)\qquad \begin{aligned} &(\ddot{u}^h(t), v)+a(u^h(t), v)=(f(t), v) \qquad \forall v\in V_h,\\ &u^h(0)=u_{0h},\ \dot{u}(0)=u_{1h}, \end{aligned}$$

where u_{0h}, $u_{1h}\in V_h$ are approximations of the initial data u_0 and u_1. This problem is equivalent to the following system of ordinary differential equations (using the notation of Section 8.2): Find $\xi(t)\in R^m$ such that

$$
(9.58)\qquad \begin{aligned} &B\ddot{\xi}+A\xi=F, \qquad t\in I,\\ &\xi(0)=\theta_0,\ \dot{\xi}(0)=\theta_1, \end{aligned}
$$

where θ_0 and θ_1 are the coordinates of u_{0h} and u_{1h} with respect to the basis $\{\varphi_1, \ldots, \varphi_M\}$ of V_h. Clearly (9.57) satisfies an energy conservation relation analogous to (9.56).

It now remains to discretize the second order system of ordinary differential equations (9.58) with respect to the time variable. To this end it is convenient to rewrite this system as follows:

$$
(9.59)\qquad \begin{aligned} &B\dot{\eta}+A\xi=F, \qquad t\in I,\\ &\dot{\xi}-\eta=0, \qquad t\in I,\\ &\xi(0)=\theta_0,\ \eta(0)=\theta_1. \end{aligned}
$$

Now, let $0<t_0<t_1, \ldots, <t_N=T$ be a subdivision of I with time steps $k_n=t_n-t_{n-1}$. We may consider the following class of time discretization methods for (9.59) (with $F=0$ for simplicity): Find $(\xi^n, \eta^n)\in R^M\times R^M$, $n=0, \ldots$, such that for $n=1, 2, \ldots N$,

$$
(9.60)\qquad \begin{aligned} &B\left(\frac{\eta^{n+1}-\eta^n}{k_n}\right)+\alpha A\xi^{n+1}+(1-\alpha)A\xi^n=0,\\ &\frac{\xi^{n+1}-\xi^n}{k_n}-(\gamma\eta^{n+1}+(1-\gamma)\eta^n)=0,\ n=1, 2, \ldots,\\ &\xi^0=\theta_0,\ \eta^0=\theta_1, \end{aligned}
$$

where $0\leqslant\alpha, \gamma\leqslant 1$ are parameters. This method is unconditionally stable for $\alpha, \gamma\geqslant\frac{1}{2}$ and second order accurate if $(\alpha, \gamma)=\left(\frac{1}{2}, \frac{1}{2}\right)$, for example. With $(\alpha, \gamma)=(0, 1)$ and a uniform subdivision in time, the scheme coincides with a well-known explicit second order centered scheme with no artificial viscosity. This particular scheme performs very well if the exact solution is smooth but not so if, for example, the initial data u_0 has a jump discontinuity, in which case severe spurious oscillations occur. With $(\alpha, \gamma)=(1, 1)$ we get a first order accurate implicit method with better stability properties but with heavy artificial viscosity. For a scheme similar to (9.60) which has been used extensively in applications, see Problem 9.19.

Prove a stability estimate for (9.60) in the case $\frac{1}{2} \leqslant \alpha \leqslant 1$, $\frac{1}{2} \leqslant \gamma \leqslant 1$.

Hint: Multiply the equations by $(\xi^{n+1}-\xi^n)$ and $B(\eta^{n+1}-\eta^n)$, respectively.

9.19 A class of time discretization methods for (9.59) well-known in the engineering literature is given by the so-called *Newmark method* (with $F=0$),

$$B\xi^{n+1}=B\xi^n+k_nB\eta^n-k_n^2(\beta A\xi^{n+1}+(\tfrac{1}{2}-\beta)\,A\xi^n),$$

$$B\eta^{n+1}=B\eta^n-k_n(\gamma A\xi^{n+1}+(1-\gamma)A\xi^n),$$

where β and γ are parameters satisfying $0\leqslant\beta\leqslant\frac{1}{2}$, $0\leqslant\gamma\leqslant 1$. Prove that this method is unconditionally stable for $2\beta\geqslant\gamma\geqslant\frac{1}{2}$. Note that with $\beta=0$ and $\gamma=\frac{1}{2}$ and a uniform subdivision, we retrieve the centered scheme obtained by taking $(\alpha, \gamma)=(0, 1)$ in (9.60).

9.20 Write the wave equation (9.54) as a first order Friedrichs' systems (cf Example 9.8).

9.21 Consider the method (9.60) with $(\alpha, \gamma)=(0, 1)$ and $k_n=k$, $n=1, 2, \ldots, N$, or equivalently the following centered scheme for (9.57): Find U^n, $n=0, 1, 2, \ldots, N$ such that for $n=1, \ldots, N-1$,

$$\frac{1}{k^2}(U^{n+1}-2U^n+U^{n-1}, v)+a(U^n, v)=(f(t_n), v) \qquad \forall v\in V_h, \tag{9.61}$$

$$U^0=u_{0h},\ U^1=U^0+ku_{1h}.$$

Prove that this scheme is conditionally stable under the condition $k\leqslant Ch$ with C sufficiently small. Hint: Rewrite (9.61) using the notation $W^n=(U^n-U^{n-1})/k$ and take $v=(W^{n+1}+W^n)/k$ to give

$$[W^{n+1}, W^{n+1}]+\frac{1}{2}\|\nabla U^{n+1}\|^2=[W^n, W^n]+\frac{1}{2}\|\nabla U^n\|^2,$$

where

$$[w, w]=(w, w)-\frac{k^2}{4}a(w, w),\ (w, w)=\|w\|^2=\|w\|^2_{L_2(\Omega)}.$$

10. Boundary element methods

10.1 Introduction

In this chapter we consider finite element methods or BEM, boundary element methods, for some integral equations arising in connection with certain elliptic boundary value problems in mechanics and physics (the presentation is based on [N], cf also [W]). As an example of such a problem let us consider the following *exterior Dirichlet problem:*

(10.1a) $\Delta u=0$ in Ω',

(10.1b) $u=u_0$ on Γ, $u(x)\rightarrow 0$ as $|x|\rightarrow\infty$,

where Ω is a bounded simply connected open set in R^3 with smooth boundary Γ, $\Omega'=R^3\backslash\bar{\Omega}$ is the complement of $\bar{\Omega}=\Omega\cup\Gamma$ (see Fig 10.1). Further $\frac{\partial}{\partial n}$ denotes differentiation in the direction n, where $n=n(x)$ is the outward unit normal to Γ at $x\in\Gamma$. This notation will be kept throughout this chapter.

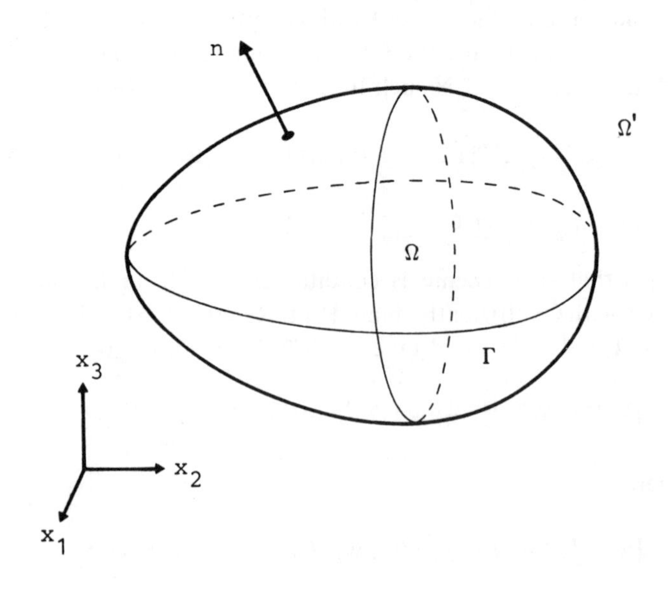

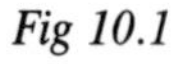
Fig 10.1

Problems of this type occur in fluid mechanics and acoustic scattering, for instance. In the latter case the equation $\triangle u=0$ is replaced by $\triangle u+\omega^2 u=0$ where ω is a given frequency. Instead of (10.1a) one may also consider homogeneous, constant coefficient elasticity or Maxwell equations corresponding to elastic or electromagnetic scattering.

Since the domain Ω' is unbounded, we cannot triangulate Ω' using a finite number of triangles, and thus we cannot apply the finite element method directly to this problem. To get a finite number of elements the first idea is simply to replace Ω' by the bounded domain $\Omega_b'=\{x\in\Omega': |x|\leqslant b\}$ for some suitably large b and use the approximate boundary condition $u(x)=0$ for $|x|=b$. To get reasonable accuracy one may have to choose b quite large, and then this procedure may be too costly.

As we will see below, since the differential equation (10.1a) is homogeneous it is possible to reformulate the problem (10.1) as an *integral equation* on the closed and bounded surface Γ. Applying a standard Galerkin or finite element method to solve this integral equation numerically, we obtain a boundary element method.

We will meet below integral equations of the following types (named after the Swedish mathematician Ivar Fredholm 1866–1927).

Fredholm equation of the first kind: Given $f: \Gamma\to R$ and the *kernel* $k: \Gamma\times\Gamma\to R$ find $q: \Gamma\to R$ such that

$$(10.2)\qquad \int_\Gamma k(x,y)q(y)d\gamma(y)=f(x), \qquad x\in\Gamma.$$

Fredholm equation of the second kind: Given $f: \Gamma\to R$ and $k: \Gamma\times\Gamma\to R$ find $q: \Gamma\to R$ such that

$$(10.3)\qquad q(x)+\int_\Gamma k(x, y)q(y)d\gamma(y)=f(x), \qquad x\in\Gamma.$$

Here $d\gamma$ is the element of surface area on Γ and $d\gamma(y)$ indicates integration with respect to the variable y. The kernels $k(x,y)$ that we will meet, will be *weakly singular;* more precisely we will have

$$(10.4)\qquad k(x,y)=\frac{c(x, y)}{|x-y|},\ x\neq y,$$

where $c(x,y)$ is a bounded function of x and y. Recall that a *singular* kernel in two dimensions (Γ is two-dimensional) behaves like $|x-y|^{-2}$ as $x\to y$. In particular, a weakly singular kernel satisfying (10.4) is integrable, i e,

$$\int_\Gamma |k(x,y)|d\gamma(y)<\infty, \qquad x\in\Gamma.$$

If we introduce the notation

$$(10.5) \qquad Kq(x)=\int_{\Gamma} k(x,y)q(y)d\gamma(y), \qquad x\in\Gamma.$$

then we may say that K is an *integral operator;* given a function q defined on Γ, a new function Kq is defined on Γ by (10.5). With this notation the problems (10.2) and (10.3) can be formulated as

$$(10.6) \qquad Kq=f \qquad \text{(Fredholm first kind)},$$

$$(10.7) \qquad (I+K)q=f \qquad \text{(Fredholm second kind)},$$

where I is the identity. In the applications below, the kernel k in (10.6) will be *symmetric,* i e, $k(x,y)=k(y,x)$, $x, y\in\Gamma$, while the kernel k in (10.7) will be non-symmetric.

10.2 Some integral equations

Let us now briefly recall some of the integral equations that arise in connection with various boundary value problems for the Laplace equation. Let us then start by recalling that the *fundamental solution* for the Laplace operator in three dimensions is given by the function

$$E(x)=-\frac{1}{4\pi|x|}.$$

By this we mean that

$$\int_{R^3} E(x)\Delta\varphi(x)dx=\varphi(0),$$

for all smooth functions φ in R^3 vanishing outside a bounded set. In other words,

$$\Delta E=\delta,$$

where δ is the delta function at 0 (cf Problem 10.2). In particular, we have that $\Delta E(x)=0$ for $x\neq 0$. Next, let us recall the following representation formula:

Theorem 10.1 If u is smooth in Ω and Ω' and

$$(10.8a) \qquad \Delta u=0 \qquad \text{in } \Omega \text{ and } \Omega',$$

(10.8b) $u(x)=0(|x|^{-1}),\ |\nabla u(x)|=0(|x|^{-2})$ as $|x|\to\infty$,

then

$$\frac{1}{4\pi}\left\{\int_\Gamma\left[\frac{\partial u}{\partial n}\right]\frac{1}{|x-y|}\,d\gamma(y)-\int_\Gamma[u]\,\frac{\partial}{\partial n_y}\left(\frac{1}{|x-y|}\right)d\gamma(y)\right\}$$

$$(10.9)\qquad =\begin{cases} u(x) & \text{if } x\notin\Gamma,\\ \dfrac{u^i(x)+u^e(x)}{2} & \text{if } x\in\Gamma,\end{cases}$$

where for $x\in\Gamma$ (i=interior, e=exterior),

$$[u]=u^i-u^e,\left[\frac{\partial u}{\partial n}\right]=\frac{\partial u^i}{\partial n}-\frac{\partial u^e}{\partial n},$$

$$u^i(x)=\lim_{\substack{y\to x\\ y\in\Omega}}u(y),\ u^e(x)=\lim_{\substack{y\to x\\ y\in\Omega'}}u(y),$$

$$\frac{\partial u^i}{\partial n}(x)=\lim_{s\to 0^-}\frac{u(x+sn)-u(x)}{s},\ \frac{\partial u^e}{\partial n}(x)=\lim_{s\to 0^+}\frac{u(x+sn)-u(x)}{s},$$

and $\dfrac{\partial}{\partial n_y}$ indicates differentiation in the direction $n(y)$.

Proof First let x be a given fixed point in Ω and define (cf Fig 10.2)

$$\Omega_\varepsilon=\{y\in\Omega:\ |y-x|>\varepsilon\},$$

$$S_\varepsilon=\{y\in R^3:\ |y-x|=\varepsilon\},$$

where ε is so small that $S_\varepsilon\subset\Omega$.

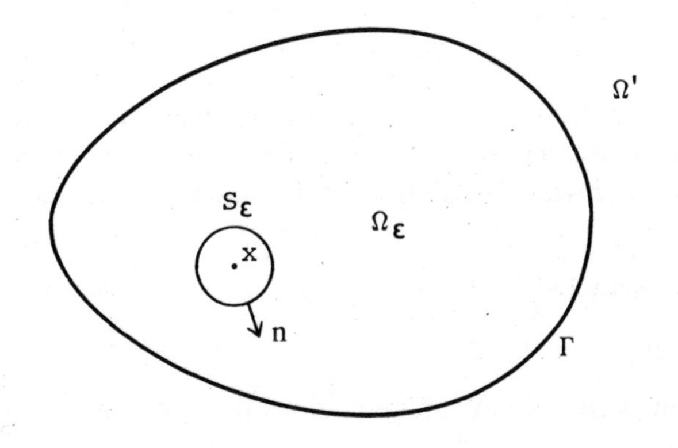

Fig 10.2

Applying Green's formula on Ω_ε to the two functions $u(y)$ and $v(y)=\frac{1}{4\pi|x-y|}$ which satisfy $\triangle u=\triangle v=0$ in Ω_ε, we get

$$(10.10)\qquad 0=\int_{\Omega_\varepsilon} u\,\triangle v\,dy-\int_{\Omega_\varepsilon} v\,\triangle u\,dy=\int_\Gamma u^i\frac{\partial v}{\partial n}\,d\gamma-\int_\Gamma v\frac{\partial u^i}{\partial n}\,d\gamma$$

$$-\int_{S_\varepsilon} u\frac{\partial v}{\partial n}\,d\gamma+\int_{S_\varepsilon} v\frac{\partial u}{\partial n}\,d\gamma,$$

where on S_ε the normal n is directed outwards (see Fig 10.2). Since u is smooth in Ω so that $\frac{\partial u}{\partial n}$ is bounded close to $x\in\Omega$, and since the area of S_ε is equal to $4\pi\varepsilon^2$, we have

$$(10.11)\qquad \left|\int_{S_\varepsilon} v\frac{\partial u}{\partial n}\,d\gamma\right|\leqslant C\int_{S_\varepsilon}\frac{d\gamma(y)}{4\pi|x-y|}=C\varepsilon\to 0 \text{ as } \varepsilon\to 0.$$

Further, for $y\in S_\varepsilon$

$$\frac{\partial v}{\partial n}(y)=-\frac{1}{4\pi|x-y|^2},$$

and thus

$$\int_{S_\varepsilon} u\frac{\partial v}{\partial n}\,d\gamma=-\frac{1}{4\pi\varepsilon^2}\int_{S_\varepsilon} u\,d\gamma\to -u(x) \text{ as } \varepsilon\to 0.$$

Hence, letting $\varepsilon\to 0$ in (10.10) we find, using similar arguments for $x\in\Gamma$ and $x\in\Omega'$, that

$$(10.12)\qquad \frac{1}{4\pi}\left\{\int_\Gamma\frac{\partial u^i}{\partial n}\frac{1}{|x-y|}d\gamma(y)-\int_\Gamma u^i\frac{\partial}{\partial n_y}\left(\frac{1}{|x-y|}\right)d\gamma(y)\right\}=\begin{cases}u(x) & x\in\Omega,\\ \frac{1}{2}u^i(x) & x\in\Gamma,\\ 0 & x\in\Omega'.\end{cases}$$

A corresponding result can be obtained by applying Green's formula on the exterior truncated domain $\Omega_b'=\{y\in\Omega': |y|\leqslant b\}$ and then letting $b\to\infty$ using (10.8b). Together with (10.12) this yields the desired representation. □

Remark The kernel $\frac{1}{|x-y|}$ is said to be a *single layer potential* and $\frac{\partial}{\partial n_y}\left(\frac{1}{|x-y|}\right)$ a *double layer potential.* □

Let us now derive the integral equations that we want to study.

10.2.1 An integral equation for an exterior Dirichlet problem using a single layer potential

Let us consider the following exterior Dirichlet problem:

$$(10.13)\qquad \begin{aligned} &\triangle u=0 && \text{in } \Omega', \\ &u=u_0 && \text{on } \Gamma, \\ &u(x)=0(|x|^{-1}),\ |\nabla u(x)|=0(|x|^{-2}), \text{ as } |x|\rightarrow \infty. \end{aligned}$$

One can show that if u_0 is sufficiently regular (more precisely, if u_0 is the restriction to Γ of some function $w\in H^1(R^3)$), then this problem has a unique solution u. We can extend this solution to the interior of Ω by letting u satisfy

$$(10.14)\qquad \begin{aligned} &\triangle u=0 && \text{in } \Omega, \\ &u=u_0 && \text{on } \Gamma. \end{aligned}$$

We know that under the condition on u_0 just stated, this problem also admits a unique solution.

By the representation formula (10.9) we now have since $[u]=0$ on Γ:

$$(10.15)\qquad u(x)=\frac{1}{4\pi}\int_\Gamma \left[\frac{\partial u}{\partial n}\right]\frac{1}{|x-y|}d\gamma(y) \qquad x\notin\Gamma,$$

and since $(u^i+u^e)/2=u_0$ on Γ,

$$u_0(x)=\frac{1}{4\pi}\int_\Gamma \left[\frac{\partial u}{\partial n}\right]\frac{1}{|x-y|}d\gamma(y), \qquad x\in\Gamma.$$

Thus, writing

$$(10.16)\qquad q=\left[\frac{\partial u}{\partial n}\right],$$

we are led to the following integral equation: Given u_0 find q such that

$$(10.17)\qquad \frac{1}{4\pi}\int_\Gamma \frac{q(y)}{|x-y|}d\gamma(y)=u_0(x), \qquad x\in\Gamma.$$

This is a Fredholm integral equation of the first kind with weakly singular kernel. Clearly the kernel is symmetric. One can show that for a large class of functions u_0, (10.17) admits a unique solution q. More precisely one can show that if $u_0\in H^s(\Gamma)$, then there exists a unique $q\in H^{s-1}(\Gamma)$ satisfying (10.17). Here and below $H^s(\Gamma)$ denotes the Sobolev space of functions defined on Γ with derivatives of order s in $L_2(\Gamma)$. With $q=\left[\frac{\partial u}{\partial n}\right]$ determined from (10.17), we obtain the solution u of (10.13) (and (10.14)) by the formula (10.15).

Another way of obtaining the integral equation (10.17) is to start out by seeking a solution of the exterior Dirichlet problem (10.13) of the form

$$u(x)=\frac{1}{4\pi}\int_\Gamma \frac{q(y)}{|x-y|}\,d\gamma(y), \qquad x\in\Omega'. \tag{10.18}$$

Since $\triangle_x\left(\frac{1}{|x-y|}\right)=0$ for $x\neq y$, where $\triangle_x$ indicates that the derivatives are taken with respect to the variable x, it is clear that a function u given by (10.18) satisfies $\triangle u=0$ in Ω'. Now letting $x\to\Gamma$ and using the fact that the right hand side of (10.18) is a continuous function of x, we obtain the integral equation (10.17) for the unknown density q.

Remark The function u defined by (10.18) for $x\in R^3$ may be interpreted for instance as the electric potential given by a distribution of electric charges on Γ with density q, or the temperature given by heat sources on Γ with intensity q. □

10.2.2 An exterior Dirichlet problem with double layer potential

When considering the exterior Dirichlet problem (10.13) let us replace (10.14) with the following *interior Neumann problem:*

$$\begin{aligned} \triangle u&=0 && \text{in } \Omega,\\ \frac{\partial u}{\partial n}\equiv\frac{\partial u^i}{\partial n}&=g && \text{on } \Gamma, \end{aligned} \tag{10.19}$$

where $g=\frac{\partial u^e}{\partial n}$. We recall that a necessary condition for (10.19) to have a solution is that

$$\int_\Gamma g\,d\gamma\equiv\int_\Gamma\frac{\partial u^e}{\partial n}\,d\gamma=0, \tag{10.20}$$

and further, a solution of (10.19) is unique only up to a constant (if u is a solution of (10.19), so is u+c for any constant c). Since $\left[\frac{\partial u}{\partial n}\right]=0$ in this case, the representation formula (10.9) gives

$$\frac{u^e(x)+u^i(x)}{2}=-\frac{1}{4\pi}\int_\Gamma [u]\frac{\partial}{\partial n_y}\left(\frac{1}{|x-y|}\right)d\gamma(y), \qquad x\in\Gamma,$$

so that

$$u^e(x) \equiv -\frac{u^i(x)-u^e(x)}{2}+\frac{u^e(x)+u^i(x)}{2}$$

$$= -\frac{\varphi}{2}-\frac{1}{4\pi}\int_\Gamma \varphi(y)\frac{\partial}{\partial n_y}\left(\frac{1}{|x-y|}\right)d\gamma(y), \qquad x\in\Gamma,$$

where

$$\varphi=[u]=u^i-u^e.$$

Thus, since $u^e(x)=u_0(x)$ for $x\in\Gamma$, we are led to the following Fredholm integral equation of the second kind:

$$\text{(10.21)} \qquad \frac{\varphi(x)}{2}+\frac{1}{4\pi}\int_\Gamma \varphi(y)\frac{\partial}{\partial n_y}\left(\frac{1}{|x-y|}\right)d\gamma(y)=-u_0(x), \quad x\in\Gamma.$$

By the representation (10.9) we also have for $x\notin\Gamma$,

$$\text{(10.22)} \qquad u(x)=-\frac{1}{4\pi}\int_\Gamma \varphi(y)\frac{\partial}{\partial n_y}\left(\frac{1}{|x-y|}\right)d\gamma(y).$$

Note that the right hand side of (10.22) is not a continuous function of x; this function has a jump equal to $\varphi=[u]$ across Γ.

To see that (10.21) is an integral equation with a weakly singular kernel of the form (10.4), we observe that for $x, y\in\Gamma$, $x\neq y$,

$$\frac{\partial}{\partial n_y}\left(\frac{1}{|x-y|}\right)=-\frac{n(y)\cdot(x-y)}{|x-y|^3}.$$

Now, if Γ is smooth, then $n(y)$ is almost orthogonal to $(x-y)$ for x close to y (see Fig 10.3). More precisely, since Γ is smooth it is easy to show that

$$n(y)\cdot(x-y)=|x-y|^2c(x,y),$$

where $c(x, y)$ is bounded function, and thus the kernel in (10.21) satisfies (10.4).

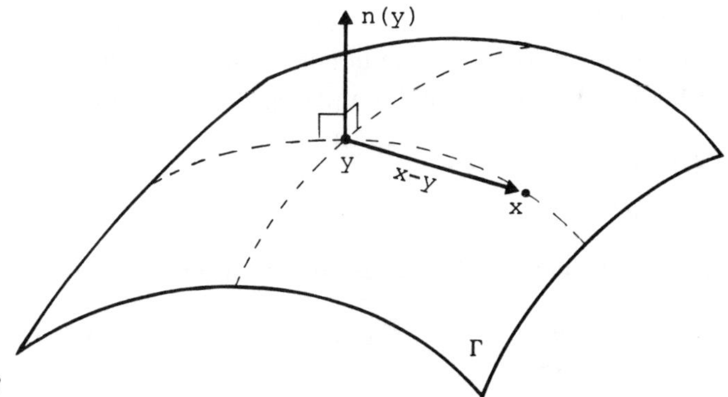

Fig 10.3

One can show that if again u_0 is the restriction to Γ of some function $w \in H^1(R^3)$ and (10.20) holds, then (10.21) admits a solution φ which is unique up to a constant.

Problems

10.1 Prove that

$$\frac{1}{4\pi}\int_\Gamma \frac{\partial}{\partial n_y}\left(\frac{1}{|x-y|}\right) d\gamma(y) = \begin{cases} -1, & x \in \Omega, \\ -\frac{1}{2}, & x \in \Gamma, \\ 0, & x \in \Omega'. \end{cases}$$

What can be said about the uniqueness of a solution of (10.21)? □

10.2 Prove using the technique of the proof of Theorem 10.1 that $E(x) = -\frac{1}{4\pi|x|}$ is a fundamental solution for the Laplace operator in R^3.

10.2.3 An exterior Neumann problem with single layer potential

Let us now consider the exterior Neumann problem

(10.23a) $\quad \Delta u = 0 \qquad$ in Ω',

(10.23b) $\quad \frac{\partial u}{\partial n} = g \qquad$ on Γ,

(10.23c) $\quad u(x) = 0(|x|^{-1}),\ |\nabla u(x)| = 0(|x|^{-2})$ as $|x| \to \infty$.

One can show that if g is sufficiently smooth (more precisely, if $g = \frac{\partial w}{\partial n}$ for some function $w \in H^1(R^3)$), then (10.23) admits a unique solution (note that for this *exterior* problem to have a solution it is not necessary that g satisfies (10.20)). With (10.23) we associate the interior Dirichlet problem:

$$\text{(10.24)} \qquad \begin{aligned} \Delta u &= 0 \qquad \text{in } \Omega, \\ u &= u^e \qquad \text{on } \Gamma. \end{aligned}$$

Since then $[u] = 0$, we have by the representation formula (10.9)

(10.25) $\quad u(x) = \frac{1}{4\pi}\int_\Gamma \frac{q(y)}{|x-y|} d\gamma(y), \qquad x \in R^3,$

where $q=\left[\frac{\partial u}{\partial n}\right]$. Now, it can be shown that if u is given by (10.25), then for $x\in\Gamma$,

$$\frac{\partial u^e}{\partial n}(x)=-\frac{q(x)}{2}+\frac{1}{4\pi}\int_\Gamma q(y)\frac{\partial}{\partial n_x}\left(\frac{1}{|x-y|}\right)d\gamma(y),$$

and using (10.23b) we are thus led to the integral equation

$$\frac{1}{2}q(x)-\frac{1}{4\pi}\int_\Gamma q(y)\frac{\partial}{\partial n_x}\left(\frac{1}{|x-y|}\right)d\gamma(y)=-g(x),\ x\in\Gamma. \tag{10.26}$$

This is again a Fredholm equation of the second kind with non-symmetric weakly singular kernel satisfying (10.4). One can show that for any $g\in L_2(\Gamma)$ there exists a unique solution $q\in L_2(\Gamma)$ of (10.26).

10.2.4 Alternative integral equation formulations

It is also possible to use a double layer potential for the exterior Neumann problem. Further, if in the representation formula (10.9) we take $u\equiv 0$ in Ω, then

$$u(x)=\frac{1}{4\pi}\left\{-\int_\Gamma \frac{\partial u}{\partial n}\frac{1}{|x-y|}d\gamma(y)+\int_\Gamma u\frac{\partial}{\partial n_y}\left(\frac{1}{|x-y|}\right)d\gamma(y)\right\}\quad \text{for } x\in\Omega' \tag{10.27}$$

where $u=u^e$ on Γ, and

$$\frac{1}{2}u(x)=\frac{1}{4\pi}\left\{-\int_\Gamma \frac{\partial u}{\partial n}\frac{1}{|x-y|}d\gamma(y)+\int_\Gamma u\frac{\partial}{\partial n_y}\left(\frac{1}{|x-y|}\right)d\gamma(y)\right\},\quad \text{for } x\in\Gamma. \tag{10.28}$$

Equation (10.28) gives an integral equation of the second kind for $u|_\Gamma$ if $\frac{\partial u}{\partial n}|_\Gamma$ is known, and an equation of the first kind for $\frac{\partial u}{\partial n}|_\Gamma$ if $u|_\Gamma$ is known. Thus, the exterior problems (10.13) and (10.23) can be solved by first solving for $u|_\Gamma$ or $\frac{\partial u}{\partial n}|_\Gamma$ in (10.28) and then using the representation (10.27).

10.3 Finite element methods

We shall now consider finite element methods for the numerical solution of the integral equations (10.17) and (10.26), that is, the equations

$$(10.29)\qquad \frac{1}{4\pi}\int_\Gamma \frac{q(y)}{|x-y|}d\gamma(y)=u_0(x), \qquad x\in\Gamma,$$

$$(10.30)\qquad \frac{1}{2}q(x)-\frac{1}{4\pi}\int_\Gamma q(y)\frac{\partial}{\partial n_x}\left(\frac{1}{|x-y|}\right)d\gamma(y)=-g(x), \qquad x\in\Gamma.$$

For simplicity we consider only the case of piecewise constant finite element approximations. Let $T_h=\{K_1, \ldots, K_M\}$ be a subdivision of Γ into "elements" K_i (e g "curved" triangles or rectangles, cf Fig 10.4) of diameter at most h. We introduce the finite element space

$$W_h=\{v\in L_2(\Gamma)\colon v|_{K_i} \text{ is constant, } i=1, \ldots, M\}.$$

We will use the basis $\{\psi_1, \ldots, \psi_M\}$ for W_h where each ψ_i is equal to one on K_i and vanishes on K_j for $j\neq i$.

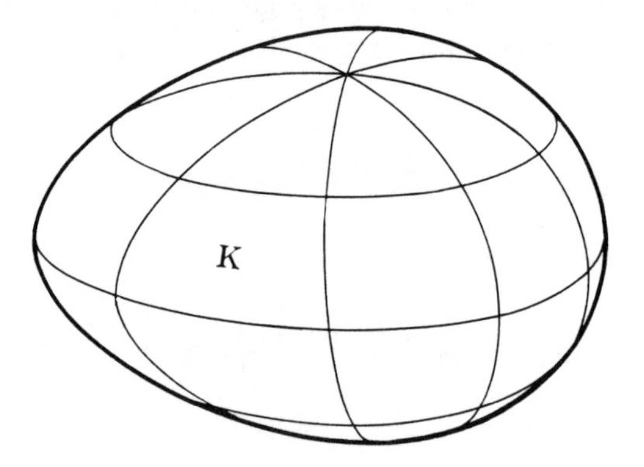

Fig 10.4

10.3.1 FEM for a Fredholm equation of the first kind

Multiplying the equation (10.29) by a test function p(x) in some test space W, as yet unspecified, and integrating over Γ, we are led to a variational formulation of (10.29) of the following form (cf [NP]): Find $q\in W$ such that

$$(10.31)\qquad b(q, p)=l(p) \qquad \forall p\in W,$$

where

$$b(q, p)=\frac{1}{4\pi}\int_\Gamma\int_\Gamma \frac{q(y)p(x)}{|x-y|}\,d\gamma(y)d\gamma(x),$$

$$l(p)=\int_\Gamma u_0(x)p(x)d\gamma(x).$$

This leads naturally to the following Galerkin method for (10.29): Find $q^h \in W_h$ such that

(10.32) $$b(q^h, p^h)=l(p^h) \qquad \forall p^h \in W_h.$$

Using the basis $\{\psi_1, \ldots, \psi_M\}$ this relation can be formulated equivalently as the linear system of equations

(10.33) $$B\xi=l,$$

where $\xi=(\xi_1, \ldots, \xi_M)\in R^M$,

$$q^h=\sum_{j=1}^{M} \xi_j\psi_j,$$

and $B=(b_{ij})$, $l=(l_i)$ with

(10.34) $$b_{ij}=\int_{K_i}\int_{K_j}\frac{d\gamma(y)d\gamma(x)}{|x-y|},$$

$$l_i=\int_{K_i} u_0 d\gamma, \quad i, j=1, \ldots, M.$$

The form $b(.\,,.)$ is bilinear and evidently symmetric. As we shall see, b is also positive definite, (i e, $b(p, p)\geqslant 0$ with equality only if $p=0$). Thus, we may define a norm $\|\cdot\|_W$ by

(10.35) $$\|p\|_W=b(p, p)^{1/2}.$$

To see that b is positive definite we recall from Section 10.2 that if

(10.36) $$v(x)=\frac{1}{4\pi}\int_\Gamma \frac{p(y)}{|x-y|}d\gamma(y), \quad x\in R^3,$$

then

(10.37a) $$\Delta v=0 \text{ in } \Omega \text{ and } \Omega',$$

(10.37b) $$\left[\frac{\partial v}{\partial n}\right]=p \text{ and } [v]=0 \text{ on } \Gamma,$$

(10.37c) $$v(x)=0(|x|^{-1}),\ |\nabla v(x)|=0(|x|^{-2}) \text{ as } |x|\to\infty.$$

Using Green's formula it follows that

$$\int_{R^3} |\nabla v|^2 dx = \int_{\Omega} \nabla v \cdot \nabla v \, dx + \int_{\Omega'} \nabla v \cdot \nabla v \, dx$$

$$= \int_{\Gamma} v^i \frac{\partial v^i}{\partial n} d\gamma - \int_{\Gamma} v^e \frac{\partial v^e}{\partial n} d\gamma = \int_{\Gamma} vp d\gamma,$$

since $v^i = v^e$ on Γ. Thus, multiplying (10.36) by p and integrating over Γ, we have

$$b(p, p) \equiv \frac{1}{4\pi} \int_{\Gamma} \int_{\Gamma} \frac{p(y)p(x)}{|x-y|} d\gamma(x) d\gamma(y) = \int_{R^3} |\nabla v|^2 dx.$$

Hence $b(p, p) \geq 0$, and if $b(p, p) = 0$, then $\nabla v \equiv 0$ so that v is constant in R^3, and hence by (10.37b) it follows that $p \equiv 0$. Thus, b is positive definite.

It now follows immediately that (10.32) admits a unique solution $q^h \in W_h$ and by our general theory for finite element methods for elliptic problems from Chapter 2, we have

$$(10.38) \qquad \|q - q^h\|_W \leq \|q - p^h\|_W \qquad \forall p^h \in W_h.$$

It can be shown that the norm $\|\cdot\|_W$ is (slightly) weaker than the $L_2(\Gamma)$-norm (ie, $\|p\|_W \leq C\|p\|_{L_2(\Gamma)}$ and in particular we have $W_h \subset W$), and thus using the L_2-norm on the right hand side of (10.38) we find by the usual approximation theory

$$\|q - q^h\|_W \leq Ch\|q\|_{H^1(\Gamma)}.$$

Remark 10.1 For readers familiar with Sobolev spaces, let us mention that here $W = H^{-1/2}(\Gamma)$ (cf [Ad]). □

Remark 10.2 One can show that the condition number of the linear system (10.33) is $0(h^{-1})$ if the triangulation T_h is quasi-uniform (this is of course related to the fact that $\|p\|_W^2 = b(p, p) \leq C\|p\|_{L_2(\Gamma)}^2$, where the last inequality is easily proved). Thus, for realistic choices of h the system (10.33) is quite well-conditioned. This is in contrast to some other Fredholm integral equations of the first kind having smooth kernels which may be very ill-conditioned and thus difficult to solve numerically. □

Remark 10.3 A large part of the computational work will have to be spent on computing the coefficients b_{ij} defined by the double integrals (10.34). If the elements K_i and K_j are not very close, then simple one-point quadrature for each integral in (10.34) may be used. If K_i and K_j are very close, then one has to be more careful and use special quadrature rules (cf [JS1] and Problem 10.5). □

Remark 10.4 Note that the matrix B in (10.33) is *dense;* we have $b_{ij} \neq 0$ $\forall i, j$. Thus, to solve (10.33) by Gaussian elimination requires $0(M^3)$ operations. □

Problem

10.3 Estimate the number of operations needed to solve (10.33) by the conjugate gradient method.

10.3.2 FEM for a Fredholm equation of the second kind

Let us now consider the equation (10.30) which we write as

$$(10.39) \qquad (I-K)q=f,$$

where $f=-2g$, I is the identity operator and $K:L_2(\Gamma) \rightarrow L_2(\Gamma)$ is the integral operator defined by

$$(10.40) \qquad Kq(x)=\frac{1}{2\pi}\int_\Gamma q(y)\frac{\partial}{\partial n_x}\left(\frac{1}{|x-y|}\right) d\gamma(y), \quad x \in \Gamma.$$

One can prove that given $f \in L_2(\Gamma)$ this problem admits a unique solution $q \in L_2(\Gamma)$, and for some constant C independent of f

$$(10.41) \qquad ||q|| \leq C||f|| \equiv C||(I-K)q||,$$

where $||\cdot||$ denotes the $L_2(\Gamma)$-norm, i e,

$$||p||=(\int_\Gamma p^2 d\gamma)^{1/2}.$$

Let us now consider the following Galerkin method for (10.39): Find $q^h \in W_h$ such that

$$(10.42) \qquad (q^h, p^h)-(Kq^h, p^h)=(f, p^h) \qquad \forall p^h \in W_h,$$

where

$$(q, p)=\int_\Gamma qp d\gamma.$$

The relation (10.42) is equivalent to the following system of linear equations

$$(10.43) \qquad (D-B)\xi=l,$$

where

$$q^h=\sum_{j=1}^{M} \xi_j \psi_j, \ \xi=(\xi_1, \ldots, \xi_M) \in R^M,$$

$B=(b_{ij})$, $D=(d_i)$ is a diagonal matrix and $l=(l_i)$ with

$$b_{ij}=\frac{1}{2\pi}\int_{K_i}\int_{K_j}\frac{\partial}{\partial n_x}\left(\frac{1}{|x-y|}\right)d\gamma(y)d\gamma(x), \quad d_i=\int_{K_i}d\gamma,$$

$$l_i=\int_{K_i} f\, d\gamma.$$

Again, the matrix B is dense but this time *non-symmetric*. For the computation of the b_{ij} Remark 10.3 again applies.

Let us analyze (10.42) and then reformulate this equation using the following notation. Let P_h: $L_2(\Gamma)\rightarrow W_h$ be the L_2-projection defined by (cf Problem 4.8).

$$(P_h q,\ p^h)=(q,\ p^h) \qquad\qquad \forall p^h\in W_h,\ q\in L_2(\Gamma).$$

By this relation it follows that $P_h p=p$ if $p\in W_h$ and

$$(P_h q,p)=(P_h q,P_h p)=(q,P_h p) \qquad \forall p,\ q\in L_2(\Gamma).$$

Since (10.42) can be written as

$$(q^h,P_h p)-(Kq^h,P_h p)=(f,P_h p) \qquad \forall p\in L_2(\Gamma),$$

we conclude that

$$(q^h,p)-(P_h Kq^h,p)=(P_h f,p) \qquad \forall p\in L_2(\Gamma),$$

or equivalently,

$$(I-K_h)q^h=P_h f,$$

where K_h: $L_2(\Omega)\rightarrow W_h$ is defined by $K_h=P_h K$. To sum up, the continuous problem and its discrete analogue can be formulated as the following equations in $L_2(\Omega)$:

(10.44) $\qquad (I-K)q=f,$

(10.45) $\qquad (I-K_h)q^h=P_h f, \qquad K_h=P_h K.$

We now want to prove a stability estimate for (10.45). Once this has been done, we obtain uniqueness and hence also existence of a solution to (10.45) and we can directly obtain an error estimate. To prove the stability of (10.45) we shall use the stability (10.41) of the continuous problem (10.44) and the following crucial property of the integral operator K:

Proposition The operator K defined by (10.40) is *smoothing;* more precisely there is a constant C such that

(10.46) $\qquad \|Kp\|_{H^1(\Gamma)}\leq C\|p\|_{L_2(\Gamma)},\ \forall p\in L_2(\Gamma).$

Here $H^1(\Gamma)$ denotes the space of functions defined on Γ with first derivatives in $L_2(\Gamma)$. This result says that if $p \in L_2(\Gamma)$, then $Kp \in H^1(\Gamma)$, and thus by applying the integral operator K, we increase the regularity by one derivative. Such a smoothing operator is also said to be *compact*.

We will also use the following error estimate for the projection operator P_h, cf Problem 4.8,

(10.47) $\quad \|p - P_h p\| \leqslant Ch \|p\|_{H^1(\Gamma)}.$

Note that since W_h consists of piecewise constants, we have in our case

$$P_h p|_{K_i} = \int_{K_i} p \, d\gamma / \int_{K_i} d\gamma = \text{mean value of p over } K_i, \ i=1, \ldots, M.$$

Now, combining (10.46) and (10.47) we find that for $p \in L_2(\Gamma)$

$$\|(K-K_h)p\| = \|(I-P_h)Kp\| \leqslant Ch \|Kp\|_{H_1(\Gamma)} \leqslant Ch \|p\|,$$

Thus,

(10.48) $\quad \|(K-K_h)p\| \leqslant Ch \|p\|, \quad \forall p \in L_2(\Gamma),$

which may also be written as $\|K-K_h\| \leqslant Ch$ where $\|A\|$ denotes the operator norm of an operator $A: L_2(\Gamma) \to L_2(\Gamma)$, i e, $\|A\| = \sup \{\|Ap\|/\|p\|: p \in L_2(\Gamma)\}$.

We can now prove the desired stability estimate:

Lemma 10.1 There are constants C and h_0 such that for $h \leqslant h_0$ and $p \in L_2(\Gamma)$

$$\|p\| \leqslant C \|(I-K_h)p\|.$$

Proof Combining (10.41) and (10.48), we have

$$\begin{aligned} \|p\| &\leqslant C\|(I-K)p\| = C\|(I-K_h)p - (K-K_h)p\| \\ &\leqslant C\|(I-K_h)p\| + C\|(K-K_h)p\| \\ &\leqslant C\|(I-K_h)p\| + C_1 h \|p\|, \end{aligned}$$

so that

$$(1 - C_1 h)\|p\| \leqslant C\|(I-K_h)p\|,$$

and the lemma follows by choosing, for instance, $C_1 h_0 = \frac{1}{2}$. □

Using this stability estimate we easily obtain the following error estimate.

Theorem 10.2 There are constants C and h_0 such that if $q \in H^1(\Gamma)$ and $q^h \in W_h$ are the solutions of (10.44) and (10.45), then for $h \leqslant h_0$,

(10.49) $\|q-q^h\|\leqslant Ch.$

Proof Substracting (10.45) from (10.44) we get

$$(I-K_h)\,(q-q^h)=(K-K_h)q+(I-P_h)f,$$

so that by Lemma 10.1

$$\|q-q^h\|\leqslant C(\|K-K_h)q\|+\|(I-P_h)f\|).$$

The desired estimate now follows from (10.47) and (10.48). □

Remark 10.5 By Lemma 10.1 it easily follows that the condition number of the matrix $(D-B)^T(D-B)$ is bounded independently of h and thus (10.43) can be solved efficiently by, for example, the conjugate gradient method applied to the least squares form of (10.43): $(D-B)^T(D-B)\xi=(D-B)^T l$. □

Remark 10.6 In certain applications it is of interest to combine the usual finite element method and the boundary element method. This is the case, for instance, if (10.1a) is replaced by the non homogenous equation $\triangle u=f$ in Ω', where we assume that the support of f is bounded so that for some $b>0$, $f(x)=0$ for $|x|\geqslant b$. We also assume that $\Omega\subset\{x:|x|\leqslant b\}$. The resulting problem may be discretized using a standard finite element method on the bounded domain $\Omega'_b=\{x\in\Omega':\ |x|\leqslant b\}$, together with an integral equation on the surface $\Gamma_b=\{x:\ |x|=b\}$ which connects the unknown values of u and the normal derivative $\frac{\partial u}{\partial n}$ on Γ_b. In this method finite elements are thus used to discretize the bounded domain Ω'_b where $f\not\equiv 0$, and a boundary integral method is used to handle the unbounded region $\{x\in\Omega':\ |x|>b\}$ where $f\equiv 0$. For more information on this topic, see [JN]. □

Problems

10.4 Consider the integral operator $K:L_2(I)\rightarrow L_2(I)$, $I=(0,1)$, defined by

$$Kq(x)=\int_I k(x,y)q(y)dy,\quad x\in I,$$

where $k(x,y)=1$ if $y\in x$ and $k(x,y)=0$ if $y>x$. Prove that

$$\|Kq\|_{H^1(I)}\leqslant C\|q\|_{L_2(I)},\ q\in L_2(I).$$

10.5 Consider the following integral with weakly singular integrand:

$$(10.50)\quad \int_K \frac{g(x,y)}{(x^2+y^2)^{1/2}}dxdy,$$

where K is the triangle with vertices (0, 0), (1, 0) and (1, 1) and g is a smooth function. Prove that the change of variable $y=xz$ transforms the integral (10.50) into

$$\int_0^1\int_0^1 \frac{g(x, xz)}{\sqrt{1+z^2}}\,dxdz. \tag{10.51}$$

This integral has a smooth integrand and may be computed using standard numerical quadrature with few quadrature points, whereas the same approach for the weakly singular integral (10.50) gives poor results. This "trick" may be used to compute the elements b_{ij} in e g (10.34) when K_i and K_j are close (in fact here the integrals with respect to x may be replaced by numerical quadrature with quadrature points at the nodes of the triangulation which leads to integrals of the form (10.50) to be calculated, cf [JS1]).

11. Mixed finite element methods

11.1 Introduction

In this chapter we briefly discuss so called *mixed finite element methods* which generalize the basic finite element method for elliptic problems described in Chapters 1–5. As an important example we shall focus on a mixed finite element method for the following Stokes problem in two dimensions (cf Section 5.2): Find the velocity $u=(u_1, u_2)$ and the pressure p such that

$$\text{(11.1a)} \qquad -\Delta u+\nabla p=f \qquad \text{in } \Omega,$$

$$\text{(11.1b)} \qquad \text{div } u=0 \qquad \text{in } \Omega,$$

$$\text{(11.1c)} \qquad u=0 \qquad \text{on } \Gamma,$$

where Ω is a bounded domain in R^2 with boundary Γ and $f=(f_1, f_2)$ is given (here of course (11.1a) is a vector equation, cf (5.6a)). Note that the pressure p is only determined up to a constant; if (u, p) solves (11.1) then (u, p+c) also solves (11.1) for any constant c. To obtain a unique pressure, we may for example impose the extra condition

$$\text{(11.1d)} \qquad \int_\Omega p\,dx=0.$$

Let us now give a variational formulation of (11.1) which generalizes the previous formulation (5.7). We shall seek u and p in the spaces V and H defined by

$$V=[H_0^1(\Omega)]^2=\{v=(v_1,v_2)\colon v_i\in H_0^1(\Omega),\ i=1,\ 2\},$$

$$H=\{q\in L_2(\Omega)\colon \int_\Omega q\,dx\equiv 0\}.$$

Notice that the velocity space V is here not restricted to divergence-free velocities as was the case in (5.7). Now multiplying (11.1a) by $v\in V$ and integrating by parts, and multiplying (11.1b) by $q\in H$, we are led to the following variational formulation of (11.1): Find $(u, p)\in V\times H$ such that

$$(11.2a)\qquad (\nabla u,\nabla v)-(p,\text{div } v)=(f,v) \qquad \forall v\in V,$$

$$(11.2b)\qquad (q,\text{div } u)=0 \qquad \forall q\in H,$$

where $(.,.)$ denotes L_2-inner products, so that in particular

$$(\nabla w,\nabla v)=\sum_{i=1}^{2}\int_\Omega \nabla w_i\cdot\nabla v_i dx,\quad (f,v)=\sum_{i=1}^{2}\int_\Omega f_i v_i dx.$$

A natural idea to get a discrete analogue of (11.2) is now to replace V and H by finite-dimensional subspaces V_h and H_h. This gives the following method: Find $(u_h, p_h)\in V_h\times H_h$ such that

$$(11.3a)\qquad (\nabla u_h,\nabla v)-(p_h, \text{div } v)=(f, v) \qquad \forall v\in V_h,$$

$$(11.3b)\qquad (q, \text{div } u_h)=0 \qquad \forall q\in H_h.$$

A method for (11.1) of the form (11.3) is called a *mixed* (finite element) *method;* the term mixed refers to the fact that in (11.3) we seek independent approximations of both the velocity u and the pressure p. With the formulation (11.3) we do not have to explicitly construct a finite element space of divergence free velocities as in (5.7), something which is difficult to do using low order polynomials (cf Section 5.2). Thus, the formulation (11.3) opens the possibility of working with velocities that only satisfy the zero divergence condition approximately through the discrete zero divergence condition (11.3b). However, we have to pay for this added freedom in the choice of V_h by introducing the pressure space H_h (cf Remark 11.1 below).

In order for (11.3) to be a reasonable discrete analogue of (11.2), the spaces V_h and H_h will have to be conveniently chosen; not just any combination will work. Loosely speaking, we want to choose V_h and H_h so that the resulting method is both stable and accurate. These demands are in some sense conflicting and one has to find a reasonable compromise. Below we will consider in detail one special choice of V_h and H_h for which stability is easily proved but which is not optimally accurate. We will also briefly give some methods with improved accuracy but omit the more elaborate proofs needed to prove the stability in these cases. Recently, modifications of (11.3) with additional stability have been introduced, see Problem 11.3. In these methods the spaces V_h and H_h can be chosen more independently.

Let us now return to the discrete Stokes problem (11.3) and consider first the stability of this problem. A natural stability inequality for (11.3) would be the following: There is a constant such that if $(u_h, p_h)\in V_h\times H_h$ satisfies (11.3), then

$$(11.4)\qquad \|u_h\|_1+\|p_h\|_0\leq C\|f\|_{-1},$$

where, (cf Remark 4.3)

$$||f||_{-1}=\sup_{v\in V}\frac{(f,v)}{||v||_1},\quad ||v||_1=(||v_1||^2_{H^1(\Omega)}+||v_2||^2_{H^1(\Omega)})^{1/2},$$

and $||q||_0=||q||_{L_2(\Omega)}$. The velocity estimate in (11.4) is obtained easily as follows. Taking $v=u_h$ in (11.3a) and $q=p_h$ in (11.3b) and adding the resulting equations, we get

$$||u_h||_1^2=(f,\ u_h)\leqslant||f||_{-1}||u_h||_1,$$

so that

$$||u_h||_1\leqslant C||f||_{-1}. \tag{11.5}$$

Next, we want to use the relation (11.3a), i e,

$$(p_h,\ \mathrm{div}\ v)=(\nabla u_h,\nabla v)-(f,\ v)\qquad \forall v\in V_h, \tag{11.6}$$

to conclude that

$$||p_h||_0\leqslant C(||u_h||_1+||f||_{-1}), \tag{11.7}$$

which together with (11.5) will prove the desired estimate (11.4). To be able to conclude (11.7) from (11.6), we need the following estimate: There is a positive constant c such that for all $q\in H_h$

$$\sup_{v\in V_h}\frac{(q,\ \mathrm{div}\ v)}{||v||_1}\geqslant c||q||_0. \tag{11.8}$$

It is clear that by using (11.8) we obtain the pressure control (11.7) from (11.5) and (11.6). The inequality (11.8) is called the *Babuska-Brezzi condition* (for the method (11.3)) and is the crucial inequality that will guarantee stability of the mixed method (11.3). Once (11.8) is established one can easily prove the following optimal error estimate for (11.3) (cf Problem 11.1 and [Br]):

$$||u-u_h||_1+||p-p_h||_0\leqslant C(\inf_{v\in V_h}||u-v||_1+\inf_{q\in H_h}||p-q||_0). \tag{11.9}$$

11.2 Some examples

We will now consider the problem of constructing the spaces V_h and H_h so that we will be able to verify (11.8). This is in fact easy to achieve by simply taking V_h large enough; the real challenge is to try to choose V_h nearly as small as possible and thus to find a good balance between V_h and H_h. In the first example we will have a situation where in fact the space V_h is "too big".

For simplicity we will in the examples below assume that Ω is a square with a uniform subdivision $T_h=\{K\}$ of Ω into squares K with side length h. The methods to be presented can be generalized directly to quadrilateral elements, cf Chapter 12.

Example 11.1 Let

(11.10a) $\quad V_h=\{v\in V: v|_K\in[Q_2(K)]^2, \forall K\in T_h\},$

(11.10b) $\quad H_h=\{q\in H: q|_K\in Q_0(K), \forall K\in T_h\}.$

In other words, V_h consists of continuous piecewise quadratic velocities and H_h of piecewise constants. We will subsequently verify that (11.8) holds in this case and thus we have by the error estimate (11.9):

$$||u-u_h||_1+||p-p_h||_0\leq Ch(h||u||_3+||p||_1).$$

This estimate is not optimal for the velocities, since optimality would require second order convergence. □

To prove (11.8) with the choice (11.10), we recall (see [GR]) that there is a constant C such that for all $q\in H$ there exists $v\in[H_0^1(\Omega)]^2$ such that

(11.11a) $\quad \text{div } v=q,$

(11.11b) $\quad ||v||_1\leq C||q||_0.$

Note that this result proves the validity of the following analogue of (11.8) for the continuous Stokes problem:

(11.12) $$\sup_{v\in V}\frac{(q, \text{div } v)}{||v||_1}\geq c||q||_0 \quad \forall q\in H.$$

Now, for a given $q\in H_h$, let $v\in V$ satisfy (11.11) and define $v_h\in V_h$ as the following interpolant of v:

(11.13a) $\quad v_h(P)=\tilde{v}(P) \qquad$ for P a corner of $K\in T_h$,

(11.13b) $\quad \int_S v_h ds=\int_S v ds \qquad$ for all sides S of T_h,

(11.13c) $\quad \int_K v_h dx=\int_K v\, dx \qquad$ for all $K\in T_h$,

where $\tilde{v}\in V_h$ is defined by $(\nabla(v-\tilde{v}),\nabla w)=0$ $\forall w\in V_h$. It is then easy to see that $||v_h||_1\leq C||v||_1$ (see Problem 11.4). Using Green's formula, (11.11a), the fact that $q\in H_h$ is constant on each $K\in T_h$ and (11.13b), we now have

$$\|q\|_0^2 = (q, \operatorname{div} v) = \sum_K \int_K q \operatorname{div} v \, dx$$

$$= \sum_K \int_{\partial K} q v \cdot n_K ds = \sum_K \int_{\partial K} q v_h \cdot n_K \, ds$$

$$= \sum_K \int_K q \operatorname{div} v_h \, dx = (q, \operatorname{div} v_h),$$

with n_K denoting the outward unit normal to ∂K, $K \in T_h$. But, recalling (11.11b), we have

$$\|v_h\|_1 \leqslant C\|v\|_1 \leqslant C\|q\|_0,$$

and thus

$$\|q\|_0 = \frac{(q, \operatorname{div} v_h)}{\|q\|_0} \leqslant C \frac{(q, \operatorname{div} v_h)}{\|v_h\|_1},$$

which proves that the stability estimate (11.8) is satisfied in the case (11.10). □

Example 11.2 The simplest example of a mixed method for Stokes problem is probably given by the $Q_1 - Q_0$ method where

$$V_h = \{v \in V: v|_K \in [Q_1(K)]^2, \ \forall K \in T_h\},$$

$$H_h = \{q \in H: q|_K \in Q_0(K), \ \forall K \in T_h\}.$$

This method does not satisfy the stability inequality (11.8) since the pressure space is "too rich". Despite this fact it is possible to prove the velocity estimate (see [JP1]),

$$\|u - u_h\|_1 \leqslant C \inf_{v \in V_h} \|u - v\|_1.$$

However, the pressures p_h may not converge to p. To obtain convergence one has to filter out some unstable components of the pressure by local smoothing (e g averages over groups of four neighbouring squares K, see [JP1]). □

Example 11.3 A good method based on square (or quadrilateral) elements K is given by

$$V_h = \{v \in V: v|_K \in [Q_2(K)]^2, \ \forall K \in T_h\},$$

$$H_h = \{q \in H: q|_K \in P_1(K), \ \forall K \in T_h\}.$$

This method, the Q_2–P_1 method, satisfies the stability inequality (11.8) and the spaces V_h and H_h are well-balanced. Note that the corresponding Q_2–Q_1

method suffers from pressure instabilities similar to those of the Q_1–Q_0 method discussed in the previous example. □

Remark 11.1 Introducing bases $\{\varphi_1, \ldots, \varphi_n\}$ and $\{\psi_1, \ldots, \psi_m\}$ for V_h and H_h, respectively, the discrete problem (11.3) can be written in matrix form as

$$A\xi - B\theta = F, \tag{11.14a}$$

$$B^T\xi = 0, \tag{11.14b}$$

where $A=(a_{ij})$, $B=(b_{ij})$, $F=(F_j)$, with

$$a_{ij}=a(\varphi_i,\varphi_j),\ b_{ij}=(\psi_i,\ \text{div}\ \varphi_j),\ F_j=\int_\Omega f\varphi_j dx,$$

and ξ and θ are the coordinates of u_h and p_h.

In order for θ to be uniquely determined from (11.14a), we clearly need to have dim $V_h \geqslant$ dim H_h. The same demand comes from (11.14b) since we want the space $\{v \in V_h$: (div v, q)=0 $\forall q \in H_h\}$ corresponding to $\{\eta \in R^n$: $B^T\eta=0\}$ to be rich enough to contain a good interpolant of u.

The system matrix in (11.14) is not positive definite and thus it is not so clear how to solve (11.14) iteratively in an efficient way, neither is it clear that Gaussian elimination without pivoting will work. One way out of this difficulty is to replace (11.14b) by the perturbed equation

$$\varepsilon\theta + B^T\xi = 0, \tag{11.15}$$

with ε a small positive constant (cf Problem 11.3). After elimination of θ, this leads to the following positive definite symmetric problem in the velocity variable ξ only

$$(A + \frac{1}{\varepsilon}BB^T)\xi = F. \tag{11.16}$$

The condition number of this problem increases with decreasing ε and may require double precision in Gaussian elemination (for accuracy reasons we would like to choose e g $\varepsilon=0(h^2)$ in the case of Example 11.1). □

Mixed methods may be used for problems other than the Stokes equations, e g for the elasticity equations, Maxwell's equations and the plate equations. In these cases the problem is formulated as a system of equations and the different unknowns are independently approximated. For more information on mixed methods we refer to [Br], [GR].

Problems

11.1 Prove the error estimate (11.9) for the method (11.3) under the assumption (11.8). *Hint:* Write (11.3) in the form: Find

$$\varphi_h=(u_h, p_h)\in\Phi_h\equiv V_h\times H_h$$

such that

$$B(\varphi_h, \psi)=L(\psi), \ \forall\psi\in\Phi_h,$$

where

$$B(\theta, \psi)=(\nabla w, \nabla v)-(r, \text{div } v)+(q, \text{div } w),$$

$$L(\psi)=(f, v), \ \theta=(w, r), \ \psi=(v, q).$$

Next prove that with $\|\psi\|=[\|v\|_1^2+\|q\|_0^2]^{1/2}$,

$$\sup_{\psi\in\Phi_h}\frac{B(\theta, \psi)}{\|\psi\|}\geqslant c\|\theta\|, \ \forall\theta\in\Phi_h,$$

by choosing for a given $\theta=(w, r)$, the function $\psi=(w+\delta z, r)$, with $z\in V_h$ chosen so that

$$-(\text{div } z, r)\geqslant c\|r\|_0^2, \ \|z\|_1=\|r\|_0,$$

and $\delta>0$ conveniently chosen (sufficiently small).

11.2 Determine the matrix equations corresponding to Example 11.2 in the case of a uniform subdivision of the unit square and interpret the resulting method as a difference method. Verify that the space $\{q\in H_h: (q, \text{div } v)=0, \forall v\in V_h\}$ contains a pressure that alternatively takes the values ±1 on a checkerboard pattern. Is the solution of (11.3) unique in this case?

11.3 Consider the following (cf (11.15)) perturbed variant of (11.3): Find $(u_h,p_h)\in V_h\times H_h$ such that

$$(\nabla u_h, \nabla v)-(p_h, \text{div } v)=(f,v) \qquad \forall v\in V_h,$$

$$h^2(\nabla q, \nabla p_h)+(q,\text{div } u_h)=0 \qquad \forall q\in H_h,$$

where now $H_h\subset H^1(\Omega)$. Prove without using the stability condition (11.8) an error estimate for $\|u-u_h\|_1$ (cf [BP], [HFB]).

11.4 Prove that there is a constant C such that if $v_h\in V_h$ is defined by (11.13), then $\|v_h\|_1\leqslant C\|v\|_1$. Note that $\tilde{v}\in V_h$ is introduced because we cannot guarantee that $|v(P)|\leqslant C\|v\|_{H^1(\Omega)}$.

12. Curved elements and numerical integration

12.1 Curved elements

In our applications of the finite element method so far we have used piecewise linear boundary approximations. For example, in the case of a two-dimensional region Ω we approximated the boundary Γ of Ω with a polygonal line. The corresponding error is of the order $0(h^2)$ where h as usual is a measure of the size of the finite elements. To achieve higher order of approximation one may approximate the boundary with piecewise polynomials of degree $k\geqslant 2$ and in this case the error due to the boundary approximation will be reduced to $0(h^{k+1})$. In a "triangulation" of the region Ω, the triangles (or quadrilaterals) close to the boundary will then have one curved side, see Fig 12.1.

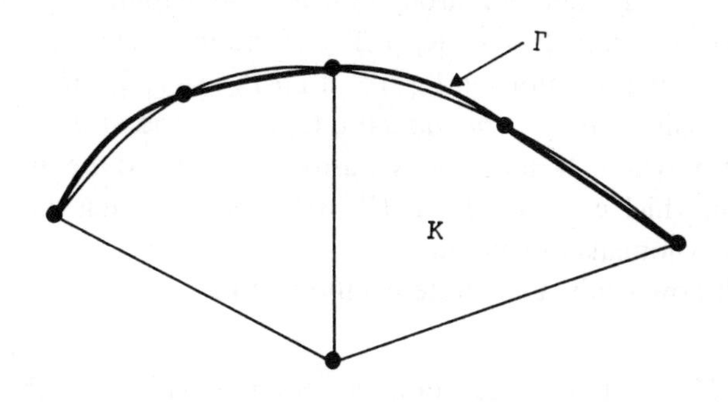

Fig 12.1

A "curved" element may be obtained in principle as follows: Suppose $(\hat{K}, P_K, \hat{\Sigma})$ is a finite element (cf Section 3.3), where $\hat{K}$ is the reference triangle in the $(\hat{x}_1, \hat{x}_2)$ – plane with vertices at $\hat{a}^1=(0,0)$, $\hat{a}^2=(1,0)$ and $\hat{a}^3=(0,1)$. For simplicity let us suppose that the degrees of freedom $\hat{\Sigma}$ are of *Lagrange type*, ie, $\hat{\Sigma}$ is a set of function values at certain points $\hat{a}^i \in \hat{K}$, $i=1, \ldots, m$. Let now F be a one-to-one mapping of $\hat{K}$ onto the "curved triangle" K in the (x_1, x_2) – plane (see Fig 10.2) with inverse F^{-1}.

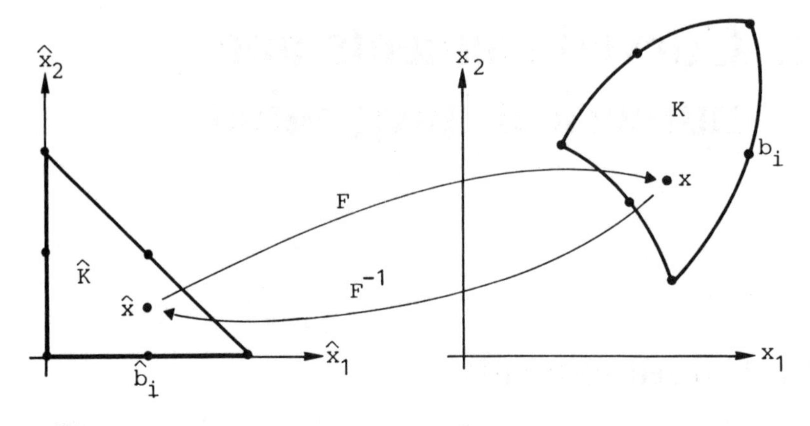

Fig 12.2

We now define

$$\text{(12.1)} \qquad P_K=\{p\colon p(x)=\hat{p}(F^{-1}(x)),\ x\in K,\ \hat{p}\in P_{\hat{K}}\},$$

$$\text{(12.2)} \qquad \Sigma_K=\{\text{the values at } a^i=F(\hat{a}^i),\ i=1, \ldots, m\}.$$

It is then easy to realize that (K, P_K, Σ_K) indeed is a finite element. The functions $p\in P_K$ are defined through the inverse mapping $F^{-1}\colon K\to\hat{K}$ and the polynomial functions $\hat{p}\colon \hat{K}\to R$, $\hat{p}\in P_K$. If the mapping $F=(F_1, F_2)$ is of the same type as the functions in P_K, i e, if $F_i\in P_K$, $i=1, 2$, then the element (K, P_K, Σ_K) is said to be of *isoparametric* type. In general the inverse mapping F^{-1} is not polynomial, unless K is a usual triangle and the midside nodes are centered in which case both F and F^{-1} are linear, and thus the functions $p\in P_K$ are not polynomials in general.

We will now study a concrete example in more detail.

Example 12.1 Let $\hat{K}$ be the reference triangle with vertices at $\hat{a}^i$, $i=1, 2, 3$ and let $\hat{a}^i$, $i=4, 5, 6$, denote the mid-points of the sides of $\hat{K}$. Further, let $P_{\hat{K}}=P_2(\hat{K})$ and let $\hat{\Sigma}$ be the values at the nodes $\hat{a}^i$, $i=1, \ldots, 6$ (cf Example 3.2) and let $\hat{\varphi}_i\in P_2(\hat{K})$, $i=1, \ldots, 6$, be the corresponding basis functions, so that $\hat{\varphi}_i(\hat{a}^j)=\delta_{ij}$. Suppose now a^i, $i=1, \ldots, 6$, are the points in the (x_1, x_2)-plane given by Fig 12.3. In particular, a^4 and a^6 are the mid points of the straight edges a^1a^2 and a^1a^3, and a^5 is slightly displaced from the straight line a^2a^3.

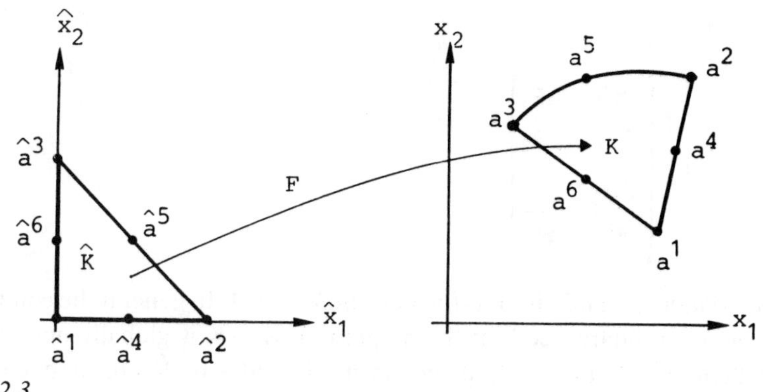

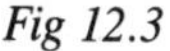
Fig 12.3

Let us now define a transformation F by

$$F(\hat{x}) = \sum_{j=1}^{6} a^j \hat{\varphi}_j(\hat{x}), \ \hat{x} \in \hat{K},$$

and let us write

$$K = F(\hat{K}) = \{x \in R^2 \colon x = F(\hat{x}), \ \hat{x} \in \hat{K}\}.$$

Then we clearly have that $F\colon\hat{K} \to K$ and $F(\hat{a}^j) = a^j$, $j = 1, \ldots, 6$, i e, the points $\hat{a}^j$ in the $\hat{x}$-plane are mapped onto the points a^j in the x-plane. We will now consider the following questions:

(a) Under what conditions is the mapping $F\colon\hat{K} \to K$ one-to-one?

(b) How can we compute the element stiffness matrix corresponding to the curved element (K, P_K, Σ_K) where P_K and Σ_K are defined by (12.1) and (12.2)?

(c) What is the interpolation error using the functions in P_K?

(d) How can we construct a finite element space V_h using the element (K, P_k, Σ_K)? Is it true that $V_h \subset C^0(\bar{\Omega})$? What is the global error?

(a) When is F one-to-one?

The mapping F is locally one-to-one in small neighbourhood of each point $\hat{x} \in \hat{K}$ if

(12.3) $\qquad \det J(\hat{x}) \neq 0, \qquad \hat{x} \in \hat{K},$

where

$$J=\begin{bmatrix} \frac{\partial F_1}{\partial \hat{x}_1} & \frac{\partial F_1}{\partial \hat{x}_2} \\ \frac{\partial F_2}{\partial \hat{x}_1} & \frac{\partial F_2}{\partial \hat{x}_2} \end{bmatrix}$$

is the *Jacobian* of F and det J is the determinant of J. In general the condition (12.3) does not guarantee that the mapping $F:\hat{K}\rightarrow K$ is globally one-to-one (cf Problem 12.1). In our case, however, the sides of $\hat{K}$ are mapped in a one-to-one fashion onto the sides of K and one can then show that (12.3) implies that $F:\hat{K}\rightarrow K$ is one-to-one, i e, for each $x\in K$ there is a unique $\hat{x}\in\hat{K}$ such that $F(\hat{x})=x$. To check if $\det J(\hat{x})\neq 0$ for $\hat{x}\in\hat{K}$, we split the transformation F in two transformations $\tilde{F}$ and $\hat{F}$,

$$F(\hat{x})=\hat{F}(\tilde{F}(\hat{x})),$$

according to the following figure:

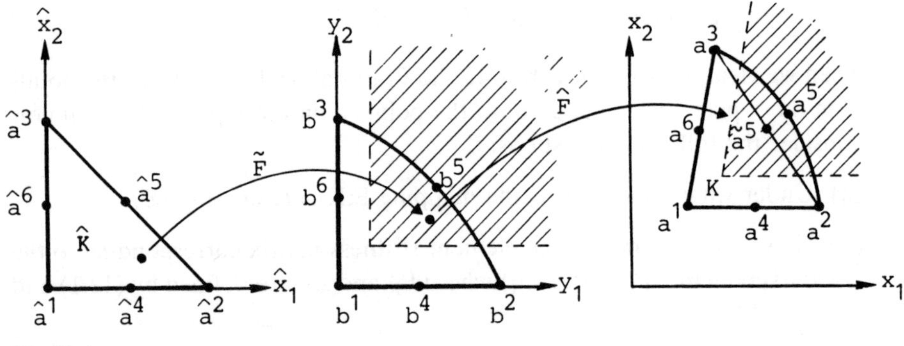

Fig 12.4

Here $\hat{F}$ is the affine mapping that maps the vertices $b^j=\hat{a}^j$, j=1, 2, 3 on the vertices $a^j=(a_1^j, a_2^j)$, j=1, 2, 3, i e,

$$\hat{F}(y)=By+b,$$

where

$$B=\begin{bmatrix} a_1^2-a_1^1 & a_1^3-a_1^1 \\ a_2^2-a_2^1 & a_2^3-a_2^1 \end{bmatrix}, \; b=\begin{bmatrix} a_1^1 \\ a_2^1 \end{bmatrix}.$$

Since by assumption the points a^1, a^2 and a^3 are not situated on the same straight line, we have that det $B \neq 0$ and $\hat{F}$ is therefore one-to-one.

It now remains to analyze the mapping $\tilde{F}$. We easily see that

$$\tilde{F}_i(\hat{x}) = \hat{x}_i + d_i\hat{x}_1\hat{x}_2, \; i=1, 2,$$

where

$$d_i = 4b_i^5 - 2.$$

The Jacobian $\tilde{J}$ corresponding to $\tilde{F}$ is given by

$$\tilde{J}(x) = \begin{bmatrix} 1+d_1\hat{x}_2 & d_1\hat{x}_1 \\ d_2\hat{x}_2 & 1+d_2\hat{x}_1 \end{bmatrix},$$

from which it follows that

$$\det \tilde{J}(\hat{x}) = 1 + d_1\hat{x}_2 + d_2\hat{x}_1,$$

and thus $\det \tilde{J}$ is linear in $\hat{x}$. Therefore $\det \tilde{J}$ is positive in $\hat{K}$ if $\det \tilde{J}$ is positive at the vertices $\hat{a}^j$, $j=1, 2, 3$. We have

$$\tilde{J}(0,0)=1, \qquad \tilde{J}(1,0)=1+d_2, \qquad \tilde{J}(0,1)=1+d_1,$$

which proves that $\det \tilde{J} > 0$ in $\tilde{K}$ if $d_i > -1$, $i=1, 2$, i e, if

$$b_i^5 > \frac{1}{4}, \quad i=1, 2.$$

We thus conclude that $\tilde{F}$ is one-to-one if b^5 and a^5 lie in the shaded areas in Fig 12.4, and thus in particular if $\tilde{a}^5$, the mid-point on the straight line a^2a^3, is close enough to a^5. Hence the original mapping F is one-to-one under the same condition. In particular, if the element K with one curved edge is used to approximate a smooth boundary, then the distance $|a^5 - \tilde{a}^5|$ (cf Fig 12.1, 12.4 and Problem 12.2) will be of the order $0(h_K^2)$, where h_K as usual is the diameter of K. Hence a^5 will be close enough to $\tilde{a}^5$ if h_k is sufficiently small, and thus we conclude that the mapping F will be one-to-one for sufficiently fine triangulations.

(b) Computation of the element stiffness matrix

The local basis functions on K are given by

$$\varphi_j(x) = \hat{\varphi}_j(F^{-1}(x)), \; j=1, \ldots, 6,$$

where $\hat{\varphi}_j$, $j=1, \ldots, 6$, is a usual basis for $P_{\hat{K}} \equiv P_2(\hat{K})$. If for example the underlying differential equation is the Poisson equation (1.16), then we have to compute the integrals

(12.4) $$a_{ij}^K=\int_K \nabla \varphi_i \cdot \nabla \varphi_j dx, \ i, j=1, \ldots, 6.$$

To this end we note that by the chain rule

$$\frac{\partial \varphi_i}{\partial x_j}=\frac{\partial}{\partial x_j}(\hat{\varphi}_i(F^{-1}(x)))=\frac{\partial \hat{\varphi}_i}{\partial \hat{x}_1}\frac{\partial \hat{x}_1}{\partial x_j}+\frac{\partial \hat{\varphi}_i}{\partial \hat{x}_2}\frac{\partial \hat{x}_2}{\partial x_j},$$

so that

$$\nabla \varphi_i=J^{-T}\nabla \hat{\varphi}_i,$$

where J^{-T} is the transposed Jacobian of the mapping F^{-1},

$$J^{-T}=\begin{bmatrix} \dfrac{\partial \hat{x}_1}{\partial x_1} & \dfrac{\partial \hat{x}_2}{\partial x_1} \\[2ex] \dfrac{\partial \hat{x}_1}{\partial x_2} & \dfrac{\partial \hat{x}_2}{\partial x_2} \end{bmatrix}.$$

If we now transform the integral in (12.4) to an integral over $\hat{K}$ using the mapping $F:\hat{K}\rightarrow K$, we get,

$$a_{ij}^K=\int_{\hat{K}}(J^{-T}\nabla \hat{\varphi}_i)\cdot(J^{-T}\nabla \hat{\varphi}_j)\ |\det J|d\hat{x}.$$

Further, by a simple calculation

$$J^{-T}=(J^{-1})^T=\frac{1}{\det J}\ J_0,$$

where

$$J_0=\begin{bmatrix} \dfrac{\partial F_2}{\partial \hat{x}_2} & -\dfrac{\partial F_2}{\partial \hat{x}_1} \\[2ex] -\dfrac{\partial F_1}{\partial \hat{x}_2} & \dfrac{\partial F_1}{\partial \hat{x}_1} \end{bmatrix},$$

so that finally

(12.5) $$a_{ij}^K=\int_{\hat{K}}(J_0\nabla \hat{\varphi}_i)\cdot(J_0\nabla \hat{\varphi}_j)\frac{d\hat{x}}{|\det J|}.$$

Thus the matrix element a_{ij}^K can be computed by evaluating an integral over the reference element $\hat{K}$. We see that the integrand is a rational function $r(\hat{x})$ (i e, $r(\hat{x})=p(\hat{x})/q(\hat{x})$ with p and q polynomials), and thus in general it is difficult to evaluate the integral (12.5) analytically. In practice this integral is most conveniently evaluated approximately by some appropriate quadrature formula, as discussed below.

(c) The interpolation error

Given a function v on K, we define the interpolant $\pi v \in P_K$ in the usual way by requiring that $\pi v(a^i)=v(a^i)$, $i=1, \ldots, 6$. If K is our usual triangle we have from Chapter 4

$$\|v-\pi v\|_{H^s(K)} \leq C h_K^{r-s} \|v\|_{H^r(K)}, \quad 0 \leq s \leq r \leq 3. \tag{12.6}$$

The estimate also holds for a curved triangle K if K is not "too curved"; more precisely if $|a^5-\tilde{a}^5|=0(h_K^2)$ with a^5 and $\tilde{a}^5$ the points of Fig 12.4. Thus, in a typical application where the curved elements approximate a smooth boundary curve, then the estimate (12.6) will hold.

(d) The corresponding space V_h

Let $T_h=\{K\}$ be a triangulation of Ω using the finite element (K, P_K, Σ_K) where the "triangles" K may have one or more curved edges. Let Ω_h be the union of the elements in T_h. Then Ω_h is an approximation of Ω with piecewise quadratic boundary (see Fig 12.1). We now define in the usual way the finite element space

$$V_h=\{v \in H^1(\Omega_h): v|_K \in P_K, K \in T_h\}.$$

It is then easy to see that if $v|_K \in P_K$ for $K \in T_h$ and v is continuous at the node points of T_h, then v is continuous across all element edges and thus $v \in H^1(\Omega_h)$. Thus, we may choose the values at the node points as global degrees of freedom. If we use this space to discretize, for example, the Poisson equation (1.16), then we have the following error estimates

$$\|u-u_h\|_{H^1(\Omega_h)} \leq C h^2 \|u\|_{H^3(\Omega)}, \tag{12.7a}$$

$$\|u-u_h\|_{L_2(\Omega_h)} \leq C h^3 \|u\|_{H^3(\Omega)}. \quad \square \tag{12.7b}$$

12.2 Numerical integration (quadrature)

We have seen above that the elements in the element stiffness matrix for a curved element contain integrals that may be difficult to evaluate exactly. We meet the same difficulty in the case of nonlinear problems or differential equations with variable coefficients. For example in the heat equation of Example 2.7 the elements of the element stiffness matrix are

$$a_{ij}^K = \int_K \sum_{m=1}^{2} k_m(x) \frac{\partial \varphi_i}{\partial x_m} \frac{\partial \varphi_j}{\partial x_m} \, dx, \tag{12.8}$$

where $k_m(x)$ is the heat conductivity in the x_m-direction at the point x.

To evaluate such integrals in practice one would use a suitably chosen numerical quadrature formula of the form:

$$(12.9) \qquad \int_K f(x)dx \sim \sum_{j=1}^{q} f(y^j)w_j$$

where the w_j, $j=1, \ldots, q$, are certain weights and the y^j are certain points in the element K.

To estimate the error committed in using the quadrature formula (12.9) we check for which polynomials p the formula (12.9) is *exact,* ie, for which polynomials p we have

$$\int_K p\,dx = \sum_{j=1} p(y^j)w_j.$$

If (12.9) is exact for $p \in P_r(K)$, then the quadrature error can be estimated as follows (if $r>0$):

$$(12.9) \qquad \left| \int_K f(x)dx - \sum_{j=1}^{n} f(y^j)w_j \right| \leqslant Ch^{r+1} \sum_{|\alpha|=r+1} \int_K |D^\alpha f|dx.$$

Let us now give some simple quadrature formulas. Here r indicates the maximal degree of the polynomials for which the formula is exact. Further, a^i $(i=1, 2, 3)$ are the vertices of the triangle K, b^j $(j=1, 2, 3)$ denote the mid points of the sides of K and a^{123} the center of gravity of K. By Q we denote a rectangle with sides parallel to the coordinate axis of lengths $2h_1$ and $2h_2$ and centered at the origin. We then have, for example, the following quadrature formulas:

$$\int_K f\,dx \sim f(a^{123})\,\text{area}(K) \qquad r=1$$

$$\int_K f\,dx \sim \sum_{j=1}^{3} f(b^j)\frac{\text{area}(K)}{3} \qquad r=2$$

$$\int_K f\,dx \sim \sum_{j=1}^{3} \left[f(a^j)\frac{\text{area}(K)}{20} + f(b^j)\frac{2\,\text{area}(K)}{15}\right] + f(a^{123})\frac{9\,\text{area}(K)}{20} \qquad r=3$$

$$\int_Q f\,dx \sim \left[f\left(\frac{h_1}{\sqrt{3}}, \frac{h_2}{\sqrt{3}}\right) + f\left(\frac{h_1}{\sqrt{3}}, -\frac{h_2}{\sqrt{3}}\right) + f\left(-\frac{h_1}{\sqrt{3}}, \frac{h_2}{\sqrt{3}}\right) + f\left(-\frac{h_1}{\sqrt{3}}, -\frac{h_2}{\sqrt{3}}\right)\right]\frac{\text{area}(Q)}{4}. \qquad r=3$$

Remark Let us return to Example 12.1 where we considered isoparametric quadratics. One can show that if the integrals (12.4) are computed with a quadrature formula which is exact for polynomials of degree $r=2$, then the error estimate (12.6) holds, that is $\|u-u_h\|_{H^1(\Omega_h)}=0(h^2)$. Further, to compute the integrals (12.8) using quadrature one should use quadrature formulas which are exact for polynomials of degree $2r-2$ when using piecewise polynomials $\varphi_i \in P_r(K)$. □

Problems

12.1 Consider the mapping $F:\hat{K}\rightarrow K$, where $F(r,\theta)=(r\sin\theta, r\cos\theta)$, $\hat{K}=\{(r,\theta): 1\leqslant r\leqslant 2, 0\leqslant\theta\leqslant 2\pi\}$, $K=\{x\in R^2: 1\leqslant|x|\leqslant 2\}$. Show that the Jacobian of F is different from zero in $\hat{K}$ and that $F:\hat{K}\rightarrow K$ is not one-to-one. This shows that the condition that the Jacobian is non-zero is not sufficient to guarantee that a mapping is globally one-to-one.

12.2 Let Γ be a circle with diameter d and let Γ_h be a polygonal approximation of Γ with vertices on Γ and with maximal side length equal to h. Prove that the maximal distance from Γ to Γ_h is of the order $h^2/4d$.

12.3 Let $\hat{K}$ be the unit square with corners $\hat{a}^i=1,\ldots,4$, let $P_K=Q_1(\hat{K})$ and let $\hat{\Sigma}$ be the degrees of freedom corresponding to the values at the $\hat{a}^i$. Prove that if K is a convex quadrilateral, then we may define an isoparametric finite element (K, P_K, Σ_K) by (12.1) and (12.2). This finite element is frequently used in applications.

13. Some non-linear problems

13.1 Introduction

In this chapter we consider some applications of finite element methods to non-linear problems in continuum mechanics. We will just indicate some aspects of this extremely rich problem area. We first consider a class of convex minimization problems generalizing the quadratic minimization problems studied in Chapters 1–7. These problems correspond to non-linear elliptic partial differential equations and so called variational inequalities. We will then discuss a non-linear parabolic problem modelling e g heat conduction with heat conduction coefficient and heat production term depending on the temperature. The finite element methods of Chapters 1 to 8 may be directly extended to these problems. Finally we consider extensions of the streamline diffusion method of Chapter 9 to the Euler and Navier-Stokes equations for an incompressible fluid, and to a model problem for compressible flow, Burgers' equation.

In all cases the discrete problems obtained after application of a finite element method, consist of non-linear systems of equations to be solved. We also comment on some iterative methods of Newton type for the numerical solution of these systems.

13.2 Convex minimization problems

13.2.1 The continuous problem

We have seen that many linear stationary problems in mechanics and physics may be formulated as minimization problems of the form

(13.1) $\quad \min_{v \in V} F(v),$

where V is a Hilbert space and $F:V \rightarrow R$ is a quadratic functional. We will now briefly discuss generalizations of (13.1) to convex functionals F that are related to non-linear elliptic problems. To formulate these generalizations we need the following definitions (as before V is a Hilbert space with scalar product $(.,.)_V$ and norm $\|\cdot\|_V$). A set $K \subset V$ is said to be *convex* if for all v, $w \in K$ and $0 \leqslant \alpha \leqslant 1$, one has

(13.2) $\alpha v+(1-\alpha)w \in K.$

The condition (13.2) states that if v, $w \in K$, then all points on the straight line between v and w also belong to K, cf Fig 13.1.

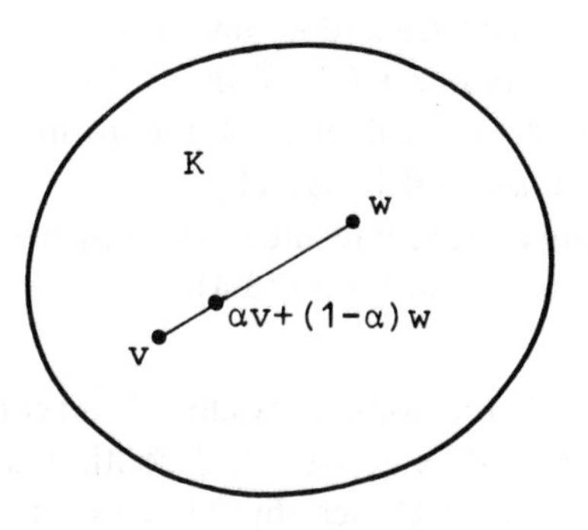

Fig 13.1

We say that a functional $F:K \rightarrow R$ defined on the convex set K is *convex* if for all v, $w \in K$ and $0 \leqslant \alpha \leqslant 1$, one has (cf Fig 13.2)

(13.3) $F(\alpha v+(1-\alpha)w) \leqslant \alpha F(v)+(1-\alpha)F(w).$

The functional F is said to be strictly convex if equality holds in (13.3) only for $\alpha=0$ or $\alpha=1$.

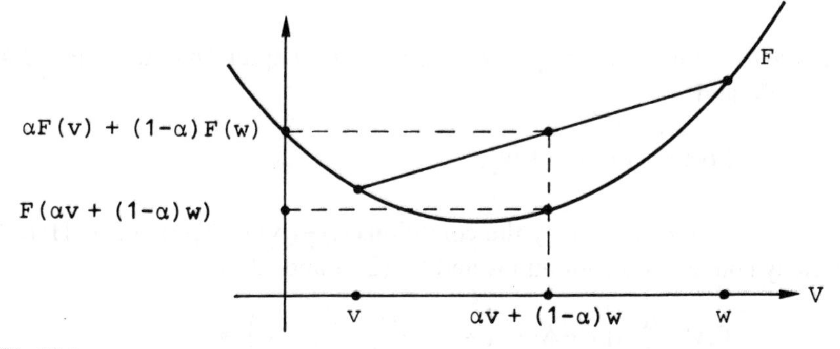

Fig 13.2

Example 13.1 A linear functional is evidently convex. The quadratic functional $F(v)=\frac{1}{2}\|v\|_V^2$, $v\in V$, is strictly convex (cf Problem 13.4). □

We further say that a set $K\subset V$ is *closed* if $x_j\in K$ and $\|x_j-x\|_V\to 0$ as $j\to\infty$ imply that $x\in K$. Finally we say that a functional $F:K\to R$ is *continuous* if $x_j\in K$ and $\|x_j-x\|_V\to 0$ as $j\to\infty$, where $x\in K$, imply that $F(x_j)\to F(x)$ as $j\to\infty$.

We consider minimization problems of the form

$$\text{(13.4)} \qquad \min_{v\in K} F(v),$$

where $K\subset V$ is a closed convex set and $F:K\to R$ is convex and continuous. If $K\neq V$, then (13.4) is a constrained minimization problem; we then seek to minimize $F(v)$ under the *side condition* $v\in K$. If $K=V$ we have an *unconstrained* minimization problem. Problems of the form (13.4) are related to variational inequalities, see [DL], [GLT].

Let us now formulate a general result concerning existence and uniqueness of solutions to problems of the form (13.4).

Theorem 13.1 Suppose K is a non-empty closed and convex set in the Hilbert space V and that $F:K\to R$ is convex and continuous. Suppose that K is bounded, i e there is constant C such that $\|v\|_V\leq C$, $\forall v\in K$, or that $F(v)\to\infty$ as $\|v\|_V\to\infty$. Then there exists a $u\in K$ such that

$$F(u)=\min_{v\in K} F(v).$$

If F is strictly convex, then u is uniquely determined.

We do not prove this result here. For a (short) proof we refer to [ET]. We now give some examples of problems in mechanics and physics that may be formulated in the form (13.4).

Example 13.2 Our standard problem (2.4) from Chapter 2 has the form (13.4) with $K=V$ and

$$F(v)=\frac{1}{2}a(v, v)-L(v),$$

where $a(.,.)$ and $L(.)$ satisfy the conditions (i)–(iv) of Section 2.1. Here F is strictly convex and continuous and by (2.2) and (2.3),

$$F(v)\geq\frac{\alpha}{2}\|v\|_V^2-\Lambda\|v\|_V=\|v\|_V\left(\frac{\alpha}{2}\|v\|_V-\Lambda\right)\to\infty$$

as $\|v\|_V \to \infty$. Thus, the assumptions of Theorem 13.1 are satisfied and existence and uniqueness of a solution of (13.4) follows in this case. Note in particular that Theorem 2.1 is essentially a special case of Theorem 13.1. □

Example 13.3 Let $V=H_0^1(\Omega)$ where Ω is a bounded domain in R^2 with boundary Γ. Let $\psi \in H^1(\Omega)$ be a given function defined on Ω such that $\psi(x) \leqslant 0$ for $x \in \Gamma$ and define

$$K=\{v \in H_0^1(\Omega):\ v \geqslant \psi \text{ in } \Omega\}.$$

Clearly K is convex and one can also show that K is closed. Further let F be defined as in Example 13.2 with a and L given by Example 2.3. Then the assumptions of Theorem 13.1 are satisfied. The unique solution $u \in K$ of (13.4) in this case represents e g the deflection of a membrane fixed at its boundary under the presence of an *obstacle* given by the function ψ. The side condition $u \in K$, i e, $u \geqslant \psi$ in Ω, corresponds to the fact that the membrane cannot penetrate the obstacle cf Fig 13.3. For more examples of similar nature, see [DL], [GLT].

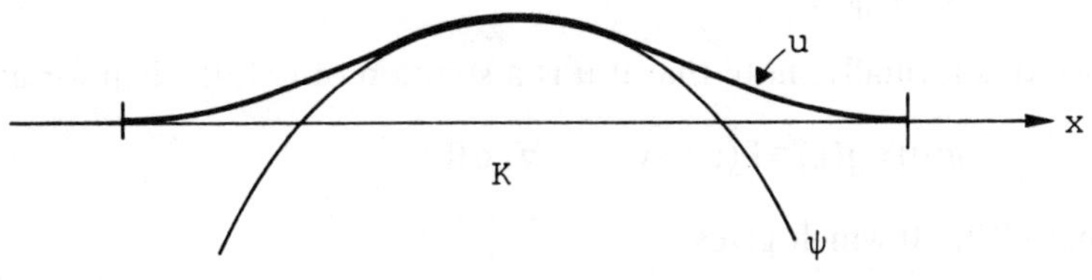

Fig 13.3

Example 13.4 Consider the non-linear elliptic problem

$$(13.5) \qquad -\frac{\partial}{\partial x_1}\left(v(|\nabla u|^2)\frac{\partial u}{\partial x_1}\right)-\frac{\partial}{\partial x_2}\left(v(|\nabla u|^2)\frac{\partial u}{\partial x_2}\right)=f \text{ in } \Omega,$$

$$u=0 \qquad \text{on } \Gamma,$$

where Ω is a two-dimensional bounded domain with boundary Γ and $v: R \to R$ is a given positive function. This problem is obtained, for example, from the following Maxwell's equations modelling a two-dimensional magnetic field problem:

$$\nabla \times H = j,$$

$$(13.6) \qquad B=\mu(|B|^2)H,$$

$$\text{div } B=0,$$

where $H=(H_1, H_2, 0)$ is the magnetic field, $B=(B_1, B_2, 0)$ is the magnetic flux density, μ is the magnetic permeability assumed to depend on $|B|^2$ and $j=(0, 0, -f)$ is a given electric current density. The problems (13.5) and (13.6) are connected through the magnetic vector potential $A=(0, 0, u)$ related to B by $B=\nabla\times A$, so that in particular $|B|^2=|\nabla u|^2$, and through the relation $\nu(\xi)=\frac{1}{\mu(\xi)}$. Now suppose that

$$(13.7) \qquad \nu(\xi)=\frac{\partial\phi}{\partial\xi}=\phi'(\xi), \qquad \xi\geqslant 0.$$

Then (13.5) corresponds to the minimization problem

$$(13.8) \qquad \operatorname*{Min}_{v\in V} F(v)$$

where

$$(13.9) \qquad F(v)=\frac{1}{2}\int_\Omega \phi(|\nabla v|^2)dx-\int_\Omega fv\ dx,$$

$$V=H_0^1(\Omega).$$

To see this formally, note that if u is a solution of (13.9), then for any $v\in V$

$$g(0)\leqslant g(\varepsilon)\equiv F(u+\varepsilon v) \qquad \forall\varepsilon\in R,$$

so that $g'(0)=0$ which gives

$$\int_\Omega \phi'(|\nabla u|^2)\nabla u\cdot\nabla v=\int_\Omega fv\ dx \qquad \forall v\in V.$$

But, using (13.7), this is a weak formulation of (13.5). Suppose now e g that ν is non-decreasing and for some positive constants ν_0 and ν_1 we have $\nu_0\leqslant\nu(\xi)\leqslant\nu_1$, $\xi\geqslant 0$. Then ϕ is convex and it is easy to see that F given by (13.9) is convex, continuous and that $F(v)\to\infty$ as $\|v\|_V\to\infty$. □

Example 13.5 Consider the minimal surface problem

$$(13.10) \qquad \operatorname*{Min}_{v=g \text{ on } \Gamma} F(v),$$

where

$$F(v)=\int_\Omega (1+|\nabla v|^2)^{1/2}\,dx,$$

where Ω is a bounded domain in the plane with boundary Γ and g is a smooth function. Here F(v) represents the area of the surface given by the graph of the function v. The solution of (13.10) corresponds to a soap film spanned by the graph of the boundary function g, cf Fig 13.4.

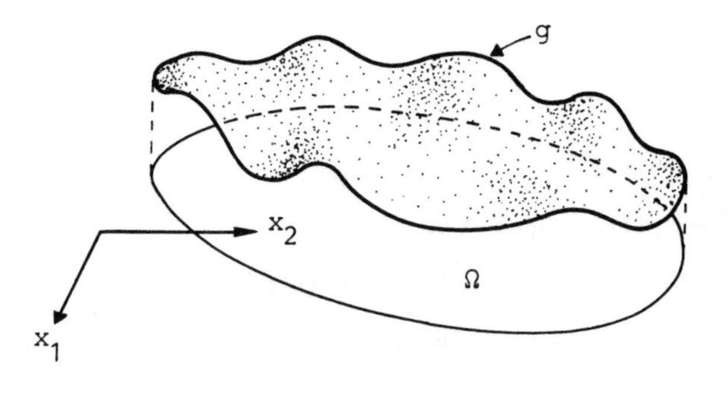

Fig 13.4

Let us now check if Theorem 13.1 may be used to prove existence of a solution of (13.10). It is natural to start with $V=H^1(\Omega)$ and $K=\{v\in H^1(\Omega)\colon v=g \text{ on } \Gamma\}$. We easily check that $F\colon K\to R$ is convex and continuous but we cannot, however guarantee that $F(v)\to\infty$ if $\|v\|_V\to\infty$, cf Problem 13.5. To be able to prove existence of a solution of (13.10), using a variant of Theorem of 13.1, the space V has to be chosen larger than $H^1(\Omega)$, (basically, V would be the space of functions on Ω whose first derivatives are integrable or more precisely, the functions on Ω of bounded variation cf [T2]). □

Example 13.6 A problem similar to (13.10) occurs as a model for the displacement of a body made of an elastic, perfectly plastic material under a load f. In two dimensions this problem takes the form (13.4) with now

$$F(v)=\int_\Omega \phi(|\nabla v(x)|)ds-\int_\Omega fv\,dx,$$

$$\phi(r)=\begin{cases}\frac{1}{2}r^2 & \text{if } 0\leq r\leq 1,\\ \frac{1}{2}-r & \text{if } r>1.\end{cases}$$

Again we need to take V larger than $H_0^1(\Omega)$ to obtain existence in general (cf Example 13.5). Also f has to be small enough to avoid having
$\inf F(v)=-\infty$, which corresponds to collapse of the elasto-plastic body and non-existence of a solution, cf [T2], [JS2]. □

Example 13.7 An alternative formulation of the elasto-plastic problem of the previous example, using stresses instead of displacements as unknowns, is as follows (cf Problem 2.9):

$$\underset{q\in K}{\text{Min}}\ \frac{1}{2}||q||^2,$$

where

$$K=H_f\cap P\subset H,$$

$$H_f=\{q\in H:\ \text{div}q+f=0 \text{ in } \Omega\},$$

$$H=\{q=(q_1,q_2):\ q_i\in L_2(\Omega)\}=[L_2(\Omega)]^2,$$

$$P=\{q\in H:\ |q(x)|\leqslant 1,\ x\in\Omega\},$$

$$||q||^2=\int_\Omega |q(x)|^2dx.$$

One can easily show that K is closed in H and the existence of a solution follows from Theorem 13.1 if K is non-empty. The latter condition will be satisfied if again the load f is below the collapse load. □

13.2.2 Discretizations

Suppose now we have a convex minimization problem of the form (13.4). A discrete analogue of this problem is obtained by replacing K with $K_h=K\cap V_h$, where V_h is a finite-dimensional subspace of V. This leads to the finite-dimensional minimization problem:

$$\underset{v\in K_h}{\text{Min}}\ F(v). \tag{13.11}$$

In Example 13.3 above we may e g choose

$$K_h=\{v\in V_h:\ v\geqslant\psi \text{ in } \Omega\},$$

where $V_h\subset H_0^1$ is a standard finite element space of piecewise linear functions on a triangulation T_h. Introducing as usual a basis $\{\varphi_1, \ldots, \varphi_M\}$ for V_h, the problem (13.11) may be written as a convex minimization problem in R^M of the form

$$\underset{\eta\in Q}{\text{Min}}\ f(\eta) \tag{13.12}$$

where $Q\subset R^M$ is a closed convex set and $f:R^M\rightarrow R$ is convex. If $Q=R^M (Q\neq R^M)$ then (13.12) is an *unconstrained (constrained) convex mini-*

mization problem in R^M. We note that the typical problem (7.1) of Chapter 7 has the form (13.12) with $Q=R^M$ and f quadratic, $f(\eta)=\frac{1}{2}\eta\cdot A\eta-b\cdot\eta$.

13.2.3 Numerical methods for convex minimization problems

Let us very briefly indicate some methods for the numerical solution of convex minimization problems of the form (13.12). We shall then use the notation of Section 7.1 with in particular f' and f'' the gradient and Hessian of f (which we assume to exist). Let us then first consider the unconstrained case with $Q=R^M$. In this case it is easy to show that $\xi\in R^M$ is a solution of (13.12), i e, $f(\xi)\leq f(\eta)\ \forall\eta\in R^M$, if and only if

(13.13) $\quad f'(\xi)=0.$

Further, if f is strictly convex, corresponding to $f''(\eta)$ being positive definite for all $\eta\in R^M$, then the solution ξ is uniquely determined.

For the numerical solution of (13.12) with $Q=R^M$ we shall consider, as in Chapter 7, iterative methods of the form

$$\xi^{k+1}=\xi^k+\alpha_k d^k,\ k=0,\ 1,\ \ldots,$$

where as earlier α_k is a step length, d^k is a search direction and ξ^0 is an initial approximation. We shall first give some examples of methods with $d^k=-H_k f'(\xi^k)$, where H_k is an $M\times M$ matrix. As in Chapter 7 we see that d^k is a descent direction if H_k is positive definite.

(a) The gradient method with optimal step length

In this method one chooses, of course,

$$d^k=-f'(\xi^k),\ \alpha_k=\alpha_k^{opt},$$

where

$$f(\xi^k+\alpha_k^{opt}\,d^k)=\min_{\alpha\geq 0} f(\xi^k+\alpha d^k).$$

As in the special case of a quadratic functional one can prove that the rate of convergence of this method is inversely proportional to the condition number of the Hessian $f''(\xi)$.

(b) Newton's method

In this method we have

$$d^k = -f''(\xi^k)^{-1} f'(\xi^k),$$

$$\alpha_k = 1.$$

Note that in this case one has to solve a linear system of equations with coefficient matrix $f''(\xi^k)$ at each step. If $f''(\xi)$ is non-singular (thus in particular if $f''(\xi)$ is positive definite) and the third derivatives of f are bounded, then Newton's method will converge quadratically (that is very quickly) in a neighbourhood of ξ. In this case there is a $\delta > 0$ and a constant C such that if $|\xi^k - \xi| \leqslant \delta$, then

$$|\xi^{k+1} - \xi| \leqslant C|\xi^k - \xi|^2.$$

The main problem with Newton's method is to get a sufficiently good initial approximation. Once this is achieved one gets the solution with very few iterations. To get such an initial approximation ξ^k one may have to choose $\alpha_k < 1$ to start with. In this case the method is a *damped Newton method.*

(c) Quasi-Newton methods

These methods are variants of Newton's method of the form

$$d^k = -H_k f'(\xi^k), \quad \alpha_k = \alpha_k^{opt},$$

where H_k is an $M \times M$ matrix which may be viewed as an approximation of $f''(\xi^k)^{-1}$. In the simplest case one may choose

$$H_k = f''(\xi^0)^{-1}, \quad k = 0, 1, \ldots,$$

which corresponds to the classical *modified Newton's method,* or one may take

$$H_k = C^{-1}, \quad k = 0, 1, \ldots$$

where $C = E^T E$ is an approximation of e g $f''(\xi^0)$ with E sparse, in which case we get preconditioned variants of the gradient method, cf Section 7.4. In a true quasi-Newton method the matrices H_k are successively updated in a simple way using the fact that the difference $f'(\xi^k) - f'(\xi^{k-1})$ gives information about $f''(\xi^k)$ and then the H_k become better approximations of $f''(\xi)^{-1}$ as k increases. The quasi-Newton methods are very efficient on large classes of problems (see e g [MS]).

(d) Generalized conjugate gradient (Fletcher-Reeves)

This method reads:

$$\xi^{k+1}=\xi^k+\alpha_k d^k,$$

$$\alpha_k=\alpha_k^{opt},$$

$$d^{k+1}=-g^{k+1}+\beta_k d^k,$$

$$\beta_k=\frac{g^{k+1}\cdot g^{k+1}}{g^k\cdot g^k},$$

where $g^k=f'(\xi^k)$ and $d^0=-g^0$. Usually a *restart* with $d^k=-g^k$ is made in this method at certain intervals, e g every M^{th} step.

Finally, let us just mention that to solve the problem (13.12) in the case $Q\neq R^M$, i e, the case of a constrained minimization problem, one may use different methods from nonlinear programming such as, for example, penalty methods, projected gradient methods, Lagrange multiplier methods, duality etc, see e g [G], [GLT].

13.3 A non-linear parabolic problem

Let us consider the following non-linear parabolic problem:

$$(13.12)\qquad \begin{aligned} &\frac{\partial u}{\partial t}-\operatorname{div}(\mu(u)\nabla u)=f(u) && \text{in } \Omega\times I,\\ &u=0 && \text{on } \Gamma\times I,\\ &u(.,0)=u_0, \end{aligned}$$

where $\mu:R\rightarrow R$ is a given function satisfying $\mu_0\leq\mu(r)\leq\mu_1$, $r\in R$ for some positive constants μ_i and $f:R\rightarrow R$ is given. This problem models heat conduction with heat conduction coefficient μ and heat production f depending on the unknown u. Systems of equations of the form (13.12) also model e g chemical reactions.

A weak formulation of (13.12) reads as follows: Find $u(t)\in V=H_0^1(\Omega)$, $t\in I$, such that

$$(13.13)\qquad \begin{aligned} &(\dot{u}(t), v)+a(u(t); u(t), v)=(f(u(t)), v) \qquad \forall v\in V,\ t\in I,\\ &u(0)=u_0, \end{aligned}$$

where

$$a(u; w, v)=\int_\Omega \mu(u)\nabla w\nabla v\, dx, \quad (v, w)=\int_\Omega vw\, dx.$$

Discretizing (13.13) by extending the backward Euler method (8.21), we get the following discrete analogue of (13.13): Find $U^n \in V_h$, $n=1, 2, \ldots, N$, such that for $n=1, \ldots, N$,

$$(13.14) \qquad (U^n-U^{n-1}, v)+k_n a(U^n; U^n, v)=k_n(f(U^n), v) \qquad \forall v \in V_h,$$

where $U^0=u_0$.

Under the assumption that f and a are globally Lipschitz continuous (i e for some constant C, $|f(r)-f(s)| \leqslant C|r-s|$ and similarly for a) the error estimate (8.42) may be generalized to (13.14) (see [EJ1]). Also the discontinuous Galerkin method (8.35) may directly be extended to (13.13) (cf Problem 13.2).

13.4 The incompressible Euler equations

13.4.1 The continuous problem

Let Ω be a simply connected bounded domain in R^2 with boundary Γ. Let us recall the Euler equations for an incompressible inviscid fluid enclosed in Ω: Given g and u_0 find the velocity $u=(u_1, u_2)$ and the pressure p such that

$$(13.15a) \qquad \dot{u}_i+u\cdot\nabla u_i+\frac{\partial p}{\partial x_i}=g_i \qquad \text{in } \Omega\times I,\ i=1, 2,$$

$$(13.15b) \qquad \operatorname{div} u=0 \qquad \text{in } \Omega\times I,$$

$$(13.15c) \qquad u\cdot n=0 \qquad \text{on } \Gamma\times I,$$

$$(15.15d) \qquad u=u_0 \qquad \text{in } \Omega \text{ for } t=0,$$

where as usual $n=n(x)$ is the outward unit normal to Γ at $x\in\Gamma$ and $I=(0, T)$. It is known that (see e g [K]) if g, u_0 and Γ are smooth and $\operatorname{div} u_0=0$ and $u_0\cdot n=0$ on Γ, then (13.15) admits a unique smooth solution for any T. The problem (13.15) is an example of a nonlinear hyperbolic problem. Note that the boundary condition $u\cdot n=0$ states that no fluid particles enter or leave the domain Ω.

In this section we will briefly consider two possible ways of discretizing (13.15) using streamline diffusion finite element methods. The first method is based on a reformulation of (13.15) using the vorticity ω and stream function ψ as variables while the second method is based on the velocity-pressure formulation (13.15).

We begin by recalling (cf Section 5.2) that since Ω is simply connected and div $u=0$ in Ω there is a unique *stream function* $\psi(x, t)$ such that for $t \in I$

$$u=\text{rot}\,\psi=\left(\frac{\partial\psi}{\partial x_2}, -\frac{\partial\psi}{\partial x_1}\right) \text{ in } \Omega,$$

$$\psi=0 \text{ on } \Gamma.$$

Alternatively ψ may be specified as the unique solution of the Poisson equation:

$$\text{(13.16a)} \qquad \begin{aligned} -\Delta\psi(.\,,t)&=\omega(.\,,t) && \text{in } \Omega,\ t\in I,\\ \psi(.\,,t)&=0 && \text{on } \Gamma,\ t\in I, \end{aligned}$$

where

$$\omega=\text{rot } u=\frac{\partial u_2}{\partial x_1}-\frac{\partial u_1}{\partial x_2}$$

is the *vorticity* of the velocity field u. Applying now the operator rot just defined to (13.15a), we obtain the following reformulation of (13.15):
Find ω: $\Omega\times I\rightarrow R$ such that

$$\text{(13.16b)} \qquad \begin{aligned} &\dot{\omega}+u(\omega)\cdot\nabla\omega=f && \text{in } \Omega\times I,\\ &\omega=\omega_0 && \text{in } \Omega \text{ for } t=0, \end{aligned}$$

where $f=\text{rot } g$, $\omega_0=\text{rot } u_0$ and $u(\omega)=\text{rot }\psi$, where ψ satisfies (13.16a). We see that (13.16b) formally has the form (9.3) with a coefficient $\beta=u(\omega)$ depending on the unknown solution ω. Notice that we do not have to specify any boundary conditions for ω in (13.16b) since by (13.15c), $u\cdot n=0$ on Γ.

We shall now indicate how to extend the streamline diffusion and the discontinuous Galerkin method of Chapter 9 to the nonlinear hyperbolic problem (13.16).

13.4.2 The streamline diffusion method in (ω, ψ)-formulation

Let $0=t_0<t_1 \ldots <t_N=T$ be a quasi-uniform subdivision of I into intervals $I_m=(t_{m-1}, t_m)$ of size h and introduce the "slabs" $S_m=\Omega\times I_m$. Let further $T_h=\{\tau\}$ be a quasi-uniform finite element triangulation of Ω with elements

τ also of size h, and introduce for a given integer $r \geq 1$ and $m=1, \ldots, N$ the finite-dimensional spaces

(13.17) $\quad U^m = \{v \in H^1(S_m): v|_K \in P_r(\tau) \times P_r(I_m), \ \forall K = \tau \times I_m, \ \tau \in T_h\},$

ie, U^m consists of continuous functions defined on S_m that are piecewise polynomial in x and polynomial in t of degree at most r. We also introduce the spaces

(13.18) $\quad \Psi^m = \{\varphi \in H^1(S_m): \varphi|_K \in P_{r+1}(\tau) \times P_r(I_m)$

$\forall K = \tau \times I_m, \ \tau \in T_h, \text{ and } \varphi = 0 \text{ on } \Gamma \times I_m\},$

ie, Ψ^m consists of continuous functions on S_m that are piecewise polynomial in x of degree $r+1$ and polynomial in t of degree r. We shall further use the following notation analogous to that of Section 9.9:

$$(w, v)^m = \int_{S_m} wv \, dxdt,$$

$$\langle w, v\rangle^m = \int_{\Omega} w(x, t_m) \, v(x, t_m) \, dx,$$

$$v_{\pm}(x, t) = \lim_{s \to 0^{\pm}} v(x, t+s), \qquad [v] = v_+ - v_-.$$

The streamline diffusion method for (13.16) can now be formulated as follows: Find $(\omega^m, \psi^m) \in U^m \times \Psi^m$, $m=1, \ldots, N$, such that for $m=1, \ldots, N$,

(13.19a) $\quad (\dot{\omega}^m + u^m(\psi^m) \cdot \nabla \omega^m, \ v + h(\dot{v} + u^m(\psi^m) \cdot \nabla v))^m$

$+ \langle[\omega^m], v_+\rangle^{m-1} = (f, \ v + h(\dot{v} + u^m(\psi^m) \cdot \nabla v))^m \qquad \forall v \in U^m,$

(13.19b) $\quad (\nabla \psi^m, \nabla \varphi)^m = (\omega^m, \varphi)^m \qquad \forall \varphi \in \Psi^m,$

where $\omega_-^0 = \omega_0$ for $t=0$ and $u^m(\psi^m) = \text{rot } \psi^m$. Here and below we also use the convention that $\omega_-^m = \omega_-^{m-1}$ for $t = t_{m-1}$.

13.4.3 The discontinuous Galerkin method in (ω, ψ)-formulation

Let $\beta: \Omega \to R^2$ be a direction field such that $\beta \cdot n$ is continuous across interelement sides of the triangulation T_h with normal directions n. We define for each $K = \tau \times I_m$, $\tau \in T_h$ with boundary ∂K:

$$\partial K_-(\beta) = \{(x, t) \in \partial K: n(x, t) \cdot \beta(x, t) + n_t(x, t) < 0\},$$

where $(n(x, t), n_t(x, t))$ denotes the outward unit normal to ∂K at $(x, t)\in\partial K$. Further let us introduce for $m=1, \ldots, N$, the space

$$W^m=\{v\in L_2(S_m): v|_K\in P_r(\tau)\times P_r(I_m), \quad \forall K=\tau\times I_m, \tau\in T_h\},$$

i e, W^m consists of possibly discontinuous functions on S_m that are piecewise polynomial in x and polynomial in t of degree at most r. Let us also for $(x, t)\in\partial K_-(\beta)$ introduce the notation

$$v_\pm(x, t)=\lim_{s\to 0^\pm} v(x+sn\cdot\beta, t+s), \; [v]=v_+-v_-.$$

We can now formulate the discontinuous Galerkin method for (13.16) as follows: Find $(\omega^m, \psi^m)\in W^m\times\Psi^m$, $m=1, \ldots, N$, such that for $m=1, \ldots, N$,

$$\sum_K \{ \int_K (\dot{\omega}^m+\beta\cdot\nabla\omega^m)v\,dxdt + \int_{\partial K_-(\beta)} [\omega^m]v_+|n\cdot\beta+n_t|ds\}=0 \qquad \forall v\in W^m,$$

$$(\nabla\psi^m, \nabla\varphi)^m = (\omega^m, \varphi)^m \qquad \forall\varphi\in\Psi^m, \tag{13.20}$$

where $\beta=\text{rot }\psi^m$, $\omega_-^0=\omega_0$ for $t=0$ and we sum over all $K=\tau\times I_m$ with $\tau\in T_h$. Note that since ψ^m is continuous in x, $\beta\cdot n=\frac{\partial\psi^m}{\partial s}$ is continuous across interelement boundaries S of T_h, where $\frac{\partial}{\partial s}$ denotes differentiation along S.

13.4.4 The streamline diffusion method in (u, p)-formulation

For $m=1, \ldots, N$, we introduce the velocity space

$$V^m=\{v: v=\text{rot }\varphi, \varphi\in\Psi^m\},$$

and the pressure space $Q^m=U^m$ where Ψ^m and U^m are given by (13.18) and (13.17). We observe that the functions v in V^m satisfy $\text{div } v=0$ in Ω, $v\cdot n=0$ on Γ and $v\cdot n$ is continuous across interelement boundaries.

We can now formulate the following streamline diffusion method for (13.15): Find $(u^m, p^m)\in V^m\times Q^m$, $m=1, \ldots, N$, such that for $m=1, \ldots, N$,

$$\sum_K \{ \int_K (\dot{u}^m+\beta\cdot\nabla u^m+\nabla p^m, v+h(\dot{v}+\beta\cdot\nabla v+\nabla q))dxdt \tag{13.21}$$

$$+ \int_{\partial K_-(\beta)} ([u^m], v_+)|n\cdot\beta+n_t|ds\} = \int_{S_m} (g, v+h(\dot{v}+\beta\cdot\nabla v+\nabla q))dxdt \qquad \forall(v, q)\in V^m\times Q^m,$$

where $\beta=u^m$, $u_-^0=u_0$ for $t=0$ and as above we sum over $K=\tau\times I_m$, $\tau\in T_h$. Further, $(.,.)$ denotes the scalar product in R^2. Note that although the velocities $v\in V^m$ satisfy the incompressibility condition div $v=0$, the pressure is still present in the formulation (13.21), cf (5.7). Note also that choosing $v=0$ in (13.21) gives the following discrete Poisson equation, with Neumann boundary condition, for the pressure p^m in terms of u^m

$$(\nabla p^m, \nabla q)=\sum_K (g-\dot{u}^m-\beta\cdot\nabla u^m, \nabla q) \qquad \forall q\in Q^m,$$

which corresponds to the following equation obtained by applying the divergence operator to (13.15a)

$$-\Delta p=-\mathrm{div}(g-\dot{u}-u\cdot\nabla u) \qquad \text{in } \Omega,$$

$$\frac{\partial p}{\partial n}=\sum_{i=1}^{2}(g_i-\dot{u}_i-u\cdot\nabla u_i)n_i \qquad \text{on } \Gamma.$$

For the methods (13.19)–(13.21) one can prove global error estimates of order $0(h^{r+\frac{1}{2}})$, see [JS].

13.5 The incompressible Navier-Stokes equations

The extension of the Euler equations (13.15) to the case of a viscous fluid with viscosity $\mu>0$, i e the Navier-Stokes equations for an incompressible fluid, reads as follows: Given g and u_0 find the velocity u and pressure p such that

(13.22a) $\quad \dot{u}_i+u\cdot\nabla u_i+\dfrac{\partial p}{\partial x_i}-\mu\Delta u_i=g_i \qquad$ in $\Omega\times I$, $i=1, 2$,

(13.22b) $\quad \mathrm{div}\, u=0 \qquad$ in $\Omega\times I$,

(13.22c) $\quad u\cdot n=u\cdot s=0 \qquad$ on $\Gamma\times I$,

(13.22d) $\quad u=u_0 \qquad$ in Ω for $t=0$,

where s is a tangential direction to Γ. We note that the boundary condition $u\cdot n=0$ in (13.15) is here supplemented by the no-slip condition $u\cdot s=0$ on Γ requiring the tangential velocity to be zero on Γ.

The Navier-Stokes equations (13.22) are an example of a non-linear mixed hyperbolic-parabolic system with nonlinear hyperbolic convection terms $u\cdot\nabla u_i$ and a linear elliptic viscous terms $-\mu\Delta u_i$. We will here be interested in the case of small viscosity μ in which case we meet the difficulties in numerical approximation discussed in Chapter 9. If μ is not small, then (13.22) is dominated by the linear viscous term and our earlier methods for Stokes problem may be directly extended to (13.22), see [GR], [T1].

To extend the streamline diffusion method of the previous subsection to the Navier-Stokes equation (13.22) with μ small we shall introduce the vorticity as an additional unknown. This is needed because the discrete velocities in the velocity space V^m are not necessarily continuous in x (the tangential velocities may be discontinuous across inter-element boundaries) and thus it is not clear how to handle the viscous term $\triangle u$. We note that if $\omega = \text{rot } u$ then since $\text{div } u = 0$, we have

$$\text{rot } \omega = \left(\frac{\partial \omega}{\partial x_2}, -\frac{\partial \omega}{\partial x_1} \right) = -\triangle u.$$

We now formulate the following streamline diffusion method for (13.22): Find $(u^m, p^m, \omega^m) \in V^m \times Q^m \times Q^m$, $m = 1, \ldots, N$, such that for $m = 1, \ldots, N$,

$$\text{(13.23a)} \quad \sum_K \{ \int_K (\dot{u}^m + \beta \cdot \nabla u^m + \nabla p, v + \delta(\dot{v} + \beta \cdot \nabla v + \nabla q))dxdt$$

$$+ \int_{\partial K_-(\beta)} ([u^m], v_+)|n \cdot \beta + n_t| ds + \mu \int_K (\text{rot } \omega^m, v + \delta(\dot{v} + \beta \cdot \nabla v + \nabla q\}dxdt$$

$$= \int_{S_m} (g, v + \delta(\dot{v} + \beta \cdot \nabla v + \nabla q))dxdt \qquad \forall (v, q) \in V^m \times Q^m,$$

$$\text{(13.23b)} \quad (\text{rot } \theta, u^m)^m = (\omega^m, \theta)^m \qquad \forall \theta \in Q^m,$$

where $\delta = \bar{C}h$ with $\bar{C}$ a sufficiently small positive constant and as above $\beta = u^m$ and $u^0_- = u_0$ for $t = 0$. Note that (13.23b) gives a discrete formulation of the relation $\omega = \text{rot } u$ together with the no-slip boundary condition $u \cdot s = 0$ on Γ.

The method (13.23) is robust, accurate, uniformly stable for $0 \leqslant \mu \leqslant h$ and suitable for complicated flows. In Example 13.8 below we present some results obtained using this method with $r = 1$, i e, piecewise linear velocity, pressure and vorticity. For an analysis of (13.23), see [JS]. Numerical results for (13.21) and (13.23) are given in [Han], cf Example 13.8.

13.6 Compressible flow: Burgers' equation

We conclude with an application of the streamline diffusion method to a model problem for compressible fluid flow, namely *Burgers' equation:* Find $u: J \times I \to R$ such that

$$\text{(13.24a)} \qquad \dot{u}+u\frac{\partial u}{\partial x}-\mu\frac{\partial^2 u}{\partial x^2}=0 \qquad (x, t)\in J\times I,$$

$$\text{(13.24b)} \qquad u(0, t)=u(1, t)=0 \qquad t\in I,$$

$$\text{(13.24c)} \qquad u(x, 0)=u_0(x), \qquad x\in J,$$

where $J=(0, 1)$, $I=(0, T)$, $\mu\geqslant 0$ and u_0 is a given initial function. If $\mu=0$ then the boundary condition $u(0, t)=0$ is enforced only if $u(0, t)\geqslant 0$ and $u(1, t)=0$ only if $u(1, t)\leqslant 0$, corresponding to inflow conditions. Let us use the notation of Section 13.4.2 with now $T_h=\{\tau\}$ a subdivision of J into subintervals τ and define

$$\mathring{U}^m=\{v\in U^m: v(x, t)=0 \text{ for } x=0, 1\}.$$

The streamline diffusion method for (13.24) with $\mu=0$ can now be formulated as follows: Find $u^m\in\mathring{U}^m$, $m=1, \ldots, N$, such that for $m=1, \ldots, N$,

$$\text{(13.25)} \qquad \left(\dot{u}^m+u^m\frac{\partial u^m}{\partial x}, v+h\left(\dot{v}+u^m\frac{\partial v}{\partial x}\right)\right)^m+\langle[u^m], v_+\rangle^{m-1}=0 \quad \forall v\in\mathring{V}^m,$$

where $u_-^0=u_0$ for $t=0$. This method can be directly extended to the case $0\leqslant\mu\leqslant h$ following Section 9.6.

One can prove (see [JSz1]) that if the finite element solutions u^m satisfying (13.25) stay uniformly bounded as the mesh size tends to zero, then the u^m will converge to the (entropy) solution u of (13.24) with $\mu=0$. For applications of streamline diffusion type methods to the compressible Euler and Navier-Stokes equations, see [HFM], [HMM], [HM1], [HM2], [JSz2], [JSzH], [Sz]. Methods of this type hold promise to be the first successful theoretically supported finite element methods for compressible flow problems with potentially extensive applications. In Example 13.9 below we give some results for (13.25) and variant of (13.25) with shock-capturing according to Remark 9.6, see [JSz1].

In Example 13.10 we give a result for a direct generalization of (13.25) to the compressible Euler equations in two space dimensions modelling gas flow in a channel with a step up.

Problems

13.1 Prove the following stability estimate for (13.25) for $m=1, \ldots, N$,

$$\|u^m(., t_m)_-\|^2_{L_2(I)}+\sum_{n=2}^{m}\|[u^n](., t_{n-1})\|^2_{L_2(I)}$$

$$+2h\sum_{n=1}^{m}\left\|\dot{u}^n+u^n\frac{\partial u^n}{\partial x}\right\|^2_{L_2(S_n)}\leqslant\|u_0\|^2_{L_2(I)}.$$

13.2 Extend the discontinuous Galerkin method (8.35) to (13.13). Prove a basic stability estimate.

13.3 Defining (U, P) on $\Omega\times(0, t_N)$ by $U|_{S_m}=u^m$, $P|_{S_m}=p^m$, $m=1, \ldots, N$, where (u^m, p^m) satisfies (13.21), prove the following stability estimate in the case $g\equiv 0$:

$$||U(., t_N)_-||^2_{L_2(\Omega)}+\sum_K\{\int_{\partial K_-(U)}|[U]|^2|n\cdot U+n_t|ds$$
$$+2h\int_K|\dot{U}+U\cdot\nabla U+\nabla P|^2dx\}\leq||u_0||^2_{L_2(\Omega)},$$

where we sum over elements $K=\tau\times I_m$, $\tau\in T_h$, $m=1, \ldots, N$.

13.4 Prove that the quadratic functional $F(v)=\frac{1}{2}||v||^2_V$ is convex, cf Example 13.1.

13.5 Prove with the notation of Example 13.5 that there is a sequence $\{v_i\}$ such that $F(v_i)\leq 1$ and $||v_i||_V\rightarrow\infty$ as $i\rightarrow\infty$.

13.6 Prove formally that the problems of Examples 13.6 and 13.7 are equivalent.

Example 13.8 In Fig 13.5 below we give for a cavity problem the velocities obtained by the method (13.23) after 5, 10 and 15 time steps with r=1 (i e piecewise linear velocities, pressure and vorticity), $\delta=h=\Delta t$, $\mu=10^{-3}$, given inlet velocity=1 and initial velocity=0 on a 8×16 mesh.

Example 13.9 In Fig 13.6 below we give the result of applying (13.25) with and without shock-capturing and with h=0.1 in a case where the exact solution of (13.24) with $\mu=0$ consists of a rarefaction wave and a shock and J is replaced by $(-\infty, \infty)$. The exact solution is represented by the dotted line (see [JSz1]).

Example 13.10 In Fig 3.7 below we give the density at time t=1.5 obtained by applying the streamline diffusion method to the compressible Euler equations in two dimensions in the case of channel flow with a step up at Mach number equal to 3 (see [JSzH]). At time t=0 the step is abruptly introduced into a uniform initial flow at Mach 3 and a shock wave is created which is reflected at the upper and lower wall of the channel. Note that the mesh is automatically refined where the exact solution is non-smooth.

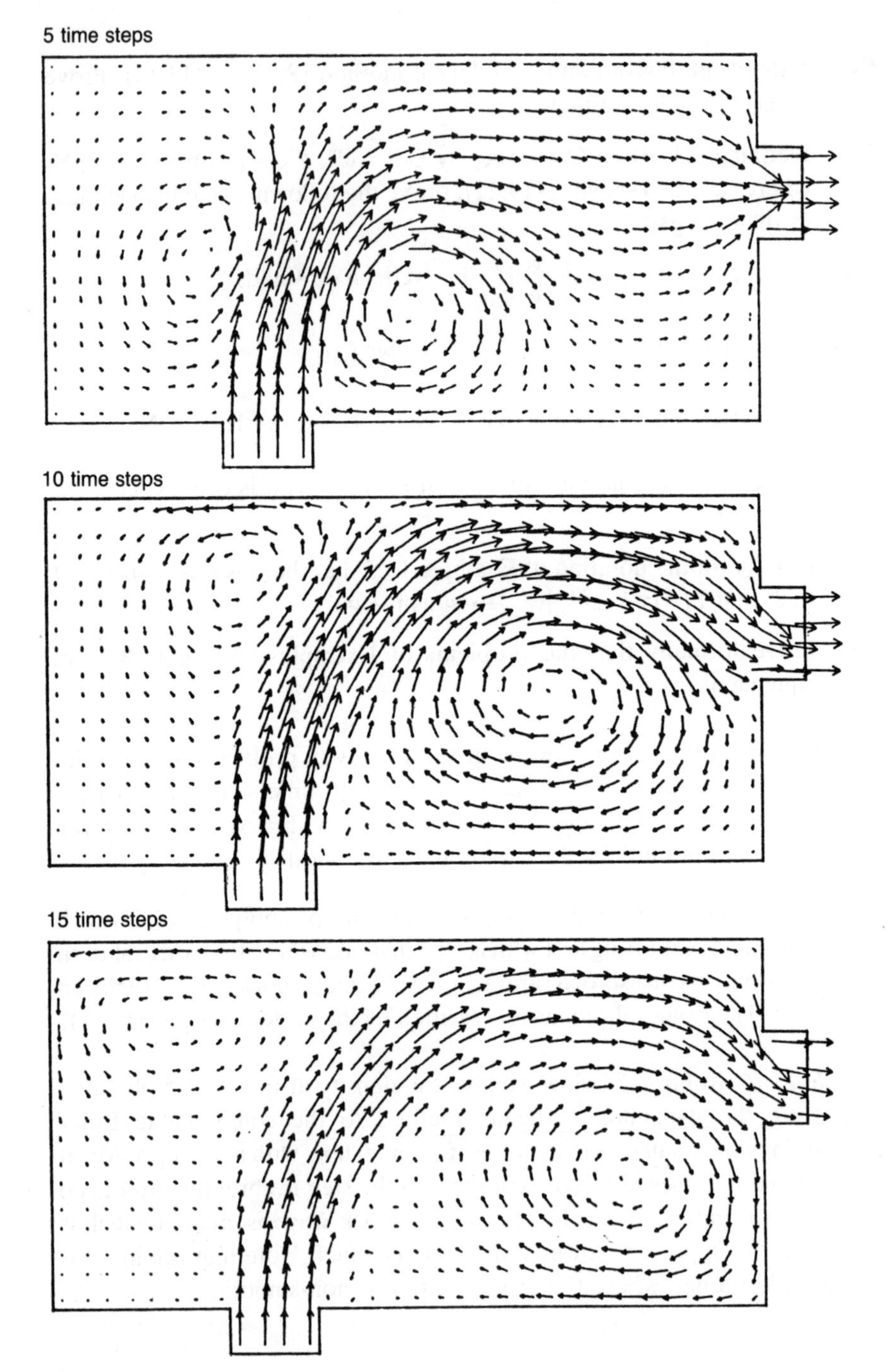

Fig 13.5 Streamline diffusion method for the incompressible Navier-Stokes equations

Fig 13.6 Streamline diffusion method for Burgers equation

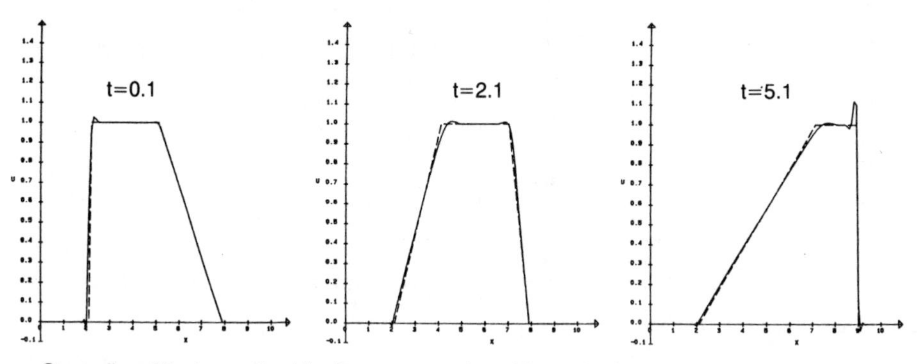

a. Streamline diffusion method for Burgers equation without shock-capturing, $\delta=h=0.1$.

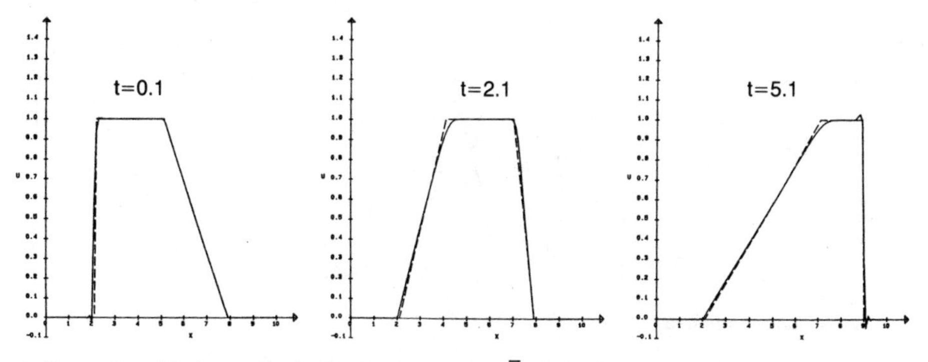

b. Streamline diffusion method with shock-capturing, $\overline{\delta}=\delta=h=0.1$.

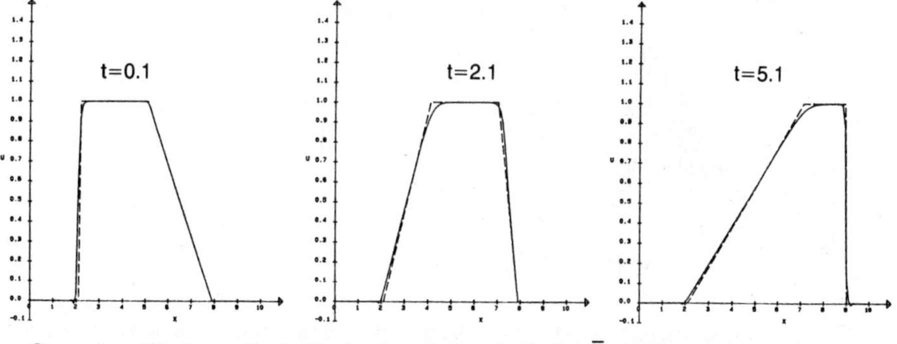

c. Streamline diffusion method with shock-capturing, $\delta=h=0.1$, $\overline{\delta}=0.25$.

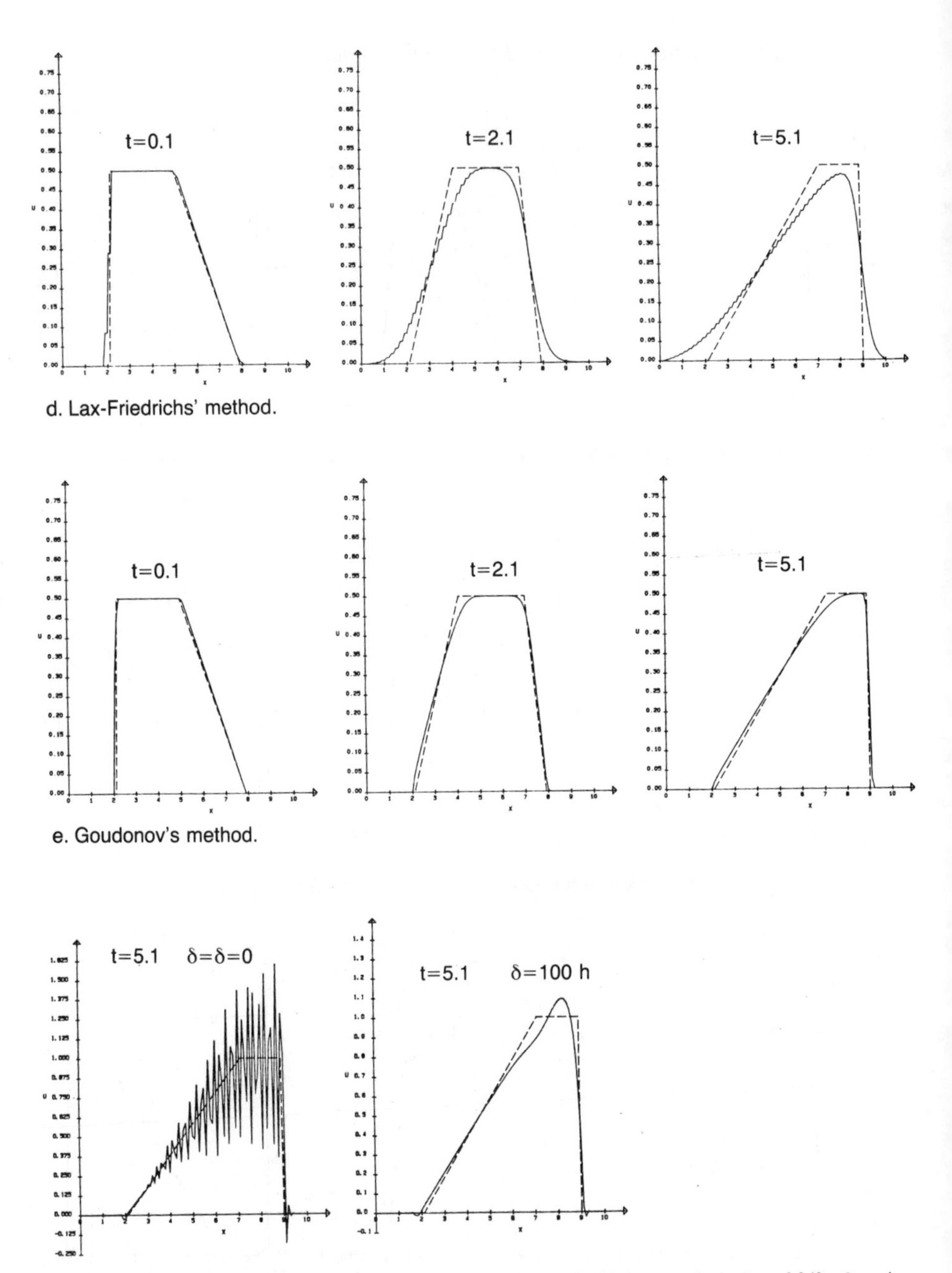

d. Lax-Friedrichs' method.

e. Goudonov's method.

f. Streamline diffusion method without shock-capturing and with incorrect choice of $\delta(\delta=0$ and $\delta=100\ h)$.

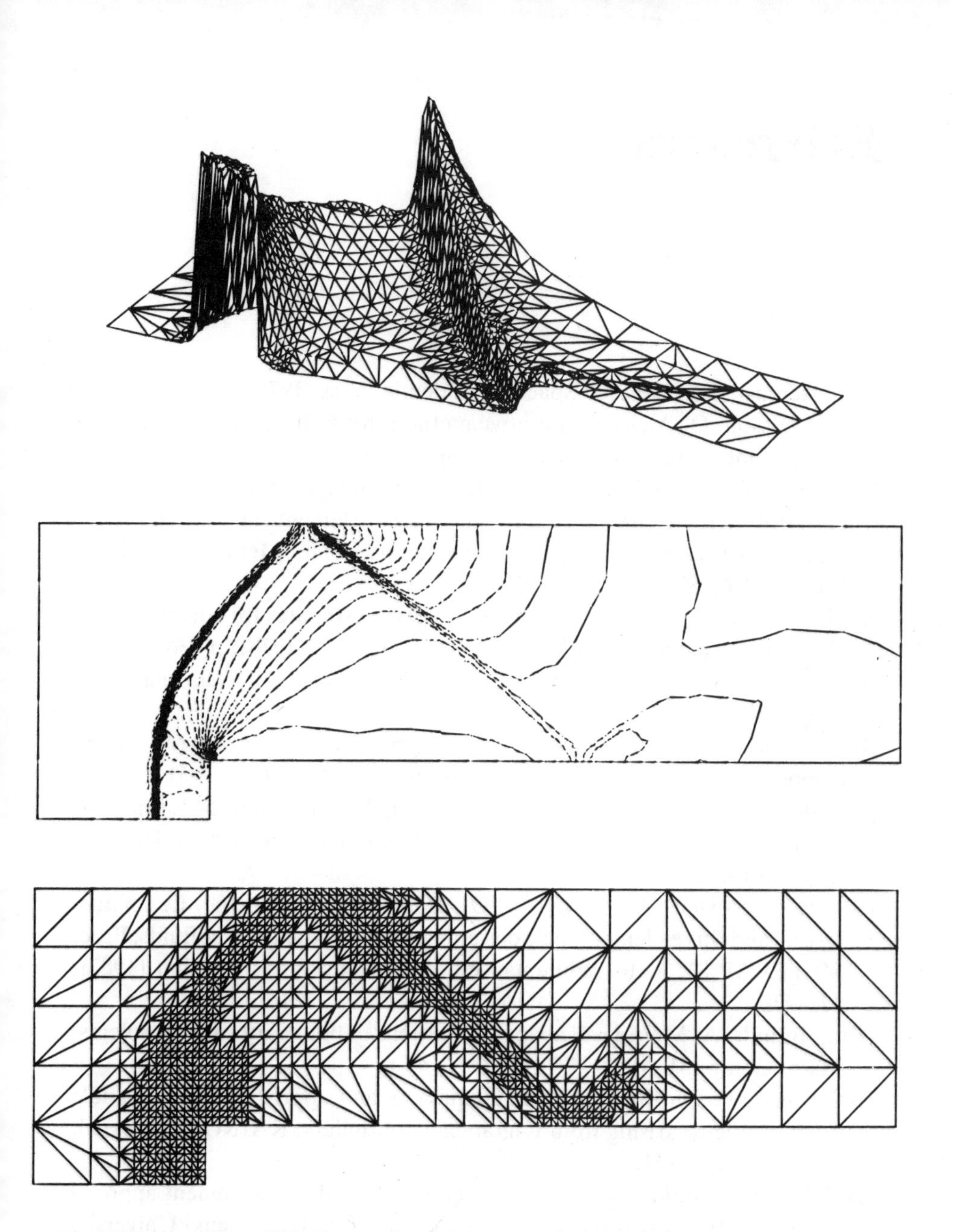

Fig 13.7 Streamline diffusion method for the compressible Euler equations for channel flow at Mach 3: surface plot and level curves for the density together with the space mesh at time $t=1.5$ after 120 time steps.

References

[Ad] Adams, R A. Sobolev Spaces, Academic Press, 1975.

[Ax] Axelsson, O. A class of iterative methods for finite element equations. Comp. Math. in Appl. Mech. and Eng. 9 (1976).

[BM] Babuska, I., and Miller, A. A feedback finite element method with a posteriori error estimates: Part I. The finite element method and some basic properties of the a posteriori estimator, Comp. Meth. Appl. Mech. Engng. 61 (1987).

[BR] Babuska, I., and Rheinboldt, W. C. Error estimates for adaptive finite element computations, SIAM J. of Num. Anal. 15(1978), p 736–754.

[Ba] Bank, R. PLTMG Users Guide, Math. Dept., Univ. of Californa at San Diego, 1981.

[BD] Bank, R. E., and Dupont, T. An optimal order process for solving elliptic finite element equations, Math. Comp. 36 (1981), 35–51.

[Be] Bercovier, M., Berold, G., and Hasbani, Y. LSD/FEM. A Library for software development in the finite element method, The Hebrew University of Jerusalem.

[BPHL] Bercovier, M., Pironneau, O., Hasbani Y., and Livne, E. Characteristic and finite element methods applied to the equations of fluids, Proc. MAFELAP 81, J. R. Whiteman ed., Academic Press, London (1982), 471–478.

[Bra] Brandt, A. Multi-level adaptive solutions to boundary value problems, Math. Comp. 31 (1977), 333–390.

[Br] Brezzi, F. On the existence, uniqueness and approximation of saddle point problems arising from Lagrangian multipliers, RAIRO Num. Anal. 8(1974), p 129–151.

[BP] Brezzi F., Pitkäranta J. On the stabilization of finite element approximations of the Stokes equations, Report MAT-A219, Helsinki University of Technology, Mathematics department, 1984.

[BH] Brooks, A., and Hughes T. Streamline upwind/Petrov-Galerkin formulations for convection dominated flows with particular emphais on the incompressible Navier-Stokes equations, Comp. Meth. Appl. Mech. Engng. 32 (1982), 199–259.

[Ci] Ciarlet, P. G. The Finite Element Method For Elliptic Problems, North Holland, 1978.

[CM] CLUB MODULEF, A library of computer procedures for finite element analysis, INRIA, Rocquencourt, France.

[DR] Douglas, J. Jr., and Russel, T. Numerical methods for convection dominated diffusion problems based on combining the method of characteristics with finite element or finite difference procedures, SIAM J. Numer. Anal. 19 (1982) 871–885.

[DS] Dupont, T., and Scott, R. Polynomial approximation of functions in Sobolev spaces, Math. Comp. 34 (1980), p 441–463.

[DL] Duvaut, G., and Lions, J. L. Les Inéquations en Mècanique et en Physique, Dunod, 1972.

[ET] Ekeland, I., and Temam, R. Convex Analysis and Variational Problems, North-Holland, 1976.

[E] Eriksson, K. Adaptive finite element methods based on optimal error estimates for linear elliptic problems, to appear in Math. Comp.

[EJ1] Eriksson, K., and Johnson, C. Error estimates and automatic time step control for non-linear parabolic problems I, SIAM J. Numer. Anal. 24 (1987), 12–22.

[EJ2] Eriksson K., and Johnson, C. An adaptive method for linear elliptic problems, Math. Comp. 50 (1988) p 361–383.

[EJL] Eriksson, K., Johnson, C., and Lennblad, J. Adaptive finite element methods for parabolic problems I: A linear model problem, to appear in SIAM J. Numer. Anal.

[EJT] Eriksson, K., Johnson, C., and Thomee, V. Time discretization of parabolic problems by the discontinuous Galerkin method, RAIRO, MAN. 19 (1985), 611–643.

[Fi] FIDAP, Fluid Dynamics Analysis Package, Fluid Dynamics International, Inc., 1600 Orrington Avenue, Suite 505, Evanston, Illinois 60201.

[F] Friedrichs, K. O. Comm. Pure and Appl. Math. 11 (1958).

[Ge] George, A. Nested dissection of regular finite element mesh, Siam J. Num. Anal. 11, 1973, 345–363.

[GR] Girault, V., and Raviart, P. A. Finite Element Approximation of the Navier-Stokes Equations. Lecture Notes in Mathematics, 749, Springer, 1979.

[G] Glowinski, R. Numerical Methods for Non-linear Variational Problems, Springer Series in Computational Physics, Springer, 1983.

[GLT] Glowinski, R., Lions, J. L., and Tremolieres, R. Numerical Analysis of Variational Inequalities, North-Holland, 1981.

[Hac] Hackbusch, W. Multigrid Methods and Applications, Springer Series in Computational Mathematics 4, Springer, 1985.

[Han] Hansbo, P. Finite element procedures for conduction and convection problems, Licenciat Thesis, Dept. of Structural Mechanics, Chalmers Univ. of Technology 1986.

[HB1] Hughes, T. J., and Brooks, A. A multidimensional upwind scheme with no crosswind diffusion, in AMD vol 34, Finite element methods for convection dominated flows, Hughes T. J. (ed): ASME, NY 1979.

[HB2] Hughes, T. J., and Brooks, A. A theoretical framework for Petrov-Galerkin methods with discontinuous weighting functions. Applications to the streamline-upwind procedure. Finite elements in fluids vol 4, ed. Gallagher, Wiley 1982.

[HFB] Hughes, T. J., Franca L., and Balestra M. A new finite element formulation for computational fluid dynamics: V. Circumventing the Babuska-Brezzi condition: A stable Petrov-Galerkin formulation of the Stokes problem accomodating equal-order interpolations, to appear in Comp. Meth. Appl. Mech. Engng.

[HFM] Hughes, T. J., Franca, L., and Mallet, M. A new finite element formulation for computational fluid dynamics: I. Symmetric forms of the compressible Euler and Navier-Stokes equations and the second law of thermodynamics, Comp. Meth. in Appl. Mech. Eng. 54 (1986), 223–234.

[HM1] Hughes, T. J., and Mallet M. A new finite element formulation for computational fluid dynamics: III. The general streamline operator for multidimensional advective-diffusive systems, Comp. Meth. Appl. Mech. 58 (1986), 305–328.

[HM2] Hughes, T. J., and Mallet M. A new finite element formulation for computational fluid dynamics: IV. A discontinuity-capturing operator for multidimensional advective-diffusive systems, Comp. Meth. Appl. Mech. 58 (1986), 329–336.

[HMM] Hughes, T. J., Mallet M., and Mizukami A. A new finite element method for computational fluid dynamics: II. Beyond SUPG, Comp. Math. in Appl. Mech. 54 (1986), 341–355.

[I] Irons, B. A frontal solution program for finite element analysis, Int. J. Num. Meth. Eng. 2 (1970), p 5–32.

[J1] Johnson C. Numerical solution of partial differential equations by the finite element method, Studentlitteratur 1980 (in Swedish).

[J2] Johnson, C. Finite element methods for convection-diffusion problems, in Computing methods in applied sciences and engineering, ed R. Glowinski, J-L. Lions, North Holland 1982.

[J3] Johnson, C. Error estimates and automatic time step control for numerical methods for stiff ordinary differential equations, SIAM J. Numer. Anal. 25 (1988), p 908–926.

[J4] Johnson, C. Streamline diffusion methods for problems in fluid mechanics, in Finite Elements in Fluids VI, Wiley, 1986.

[JN] Johnson. C., and Nedelec J. C. On the coupling of boundary integral and finite element methods, Math. Comp. 35 (1980), 1063–1079.

[JNT] Johnson, C., Nie, Y. Y., and Thomée, V. An a posteriori error estimate and automatic time step control for a backward Euler discretization of a parabolic problem, Technical report 1985-23, Math. Dept., Chalmers Univ. of Technology, to appear in SIAM J. Numer. Anal.

[JN] Johnson, C., and Nävert U. An analysis of some finite element methods for advection-diffusion problems, in Axelsson, Frank, Van der Sluis (eds), Analytical and numerical approaches to asymptotic problems in analysis, North Holland, 1981.

[JNP] Johnson, C., Nävert, U., and Pitkaranta, J. Finite element methods for linear hyperbolic problems. Comp. Meth. in Appl. Mech. Engng. 45 (1984), p 285–312.

[JP1] Johnson, C., and Pitkaranta, J. Analysis of some mixed methods related to reduced integration, Math. Comp. 38 (1982), p 345–400.

[JP2] Johnson, C., and Pitkaranta, J. An analysis of the discontinuous Galerkin method for a scalar hyperbolic equation, Math. Comp. 46 (1986), p 1–26.

[JSa] Johnson, C., and Saranen, J. Streamline diffusion methods for the incompressible Euler and Navier-Stokes equations, Math. Comp. 47 (1986), 1–18.

[JSW] Johnson, C., Schatz, A., and Wahlbin, L. Crosswind smear and pointwise errors in streamline diffusion methods, Math. Comp. 49 (1987), p 25–38.

[JS1] Johnson, C., and Scott, R. An analysis of quadrature errors in second-kind boundary integral methods, to appear in SIAM J. Numer Anal.

[JS2] Johnson, C., and Scott, R. Finite element methods for plasticity problems using discontinous trial functions, in Nonlinear Finite Element Analysis in Structural Mechanics, Proc of the Europe – US Workshop Ruhr-Universität Bochum, Germany, 1980, eds. W. Wunderlich, E. Stein and K. J. Bathe, Springer 1981.

[JSz1] Johnson, C., and Szepessy, A. On the convergence of a finite element method for a nonlinear hyperbolic conservation law, Math. Comp. 49 (1987), p 427–444.

[JSz2] Johnson, C., and Szepessy, A. On the convergence of streamline diffusion finite element methods for hyperbolic conservation laws, in Numerical Methods for Compressible Flows – Finite difference, Element and Volume Techniques – AMD. Vol 78 (1987), eds. Tezduyar T., and Hughes, T. J.

[JSzH] Johnson, C., Szepessy, A. and Hansbo, P., On the convergence of shock-capturing streamline diffusion finite element methods for conservation laws, to appear in Math. Comp.

[K] Kato, T. On classical solutions of the two-dimensional non-stationary Euler equations, Arch. Rational. Mech. Anal. 25 (1967), p 186–200.

[Le] Lesaint P. Finite element methods for symmetric hyperbolic systems, Num. Math. 21 (1973), 244–255.

[LRa] Lesaint P., and Raviart, P. A. On a finite element method for solving the neutron transport equation, in Mathematical Aspects of Finite Elements in Partial Differential Equations, C. de Boor. ed., Academic Press, NY 1974.

[Lu] Luenberger. Introduction to Linear and Non-linear Programming, Addison-Wesley, 1973.

[LR] Luskin, M., and Rannacher, R. On the smoothing property of the Galerkin method for parabolic equations, SIAM J. Numer. Anal. 19 (1981), 93–113.

[MS] Matthies, H., and Strang, G. The solution of non-linear finite element equations, Int. J. Num. Meth. Eng 14 (1979), 1613–1626.

[Me] Meijernik, J. Van der Vorst, M. An iterative solution method for linear systems of which the systems matrix is a symmetric M-matrix, Math. Comp. 31 (1977).

[M] Morton, K. W., Generalized Galerkin methods for hyperbolic problems, Comp. Meth. Appl. Mech. Engng. 52 (1985), 847–871.

[Ne] Necas, J. Les méthodes Directes en Theorie des Equations Elliptiques, Masson, 1967.

[N] Nedelec J. C. Cour de l'Ecole d'Eté d'Analyse Numerique, CEA, IRIA, EPF, 1977.

[NP] Nedelec, J. C., and Planchard, J. Une méthode variationelle d'éléments finis pour la résolution numérique d'un probléme extérieur dans R^3, RAIRO R3(1973), 105–129.

[Ni] Nitsche, J. A. On Korn's second inequality, RAIRO Analyse Numérique 15 (1981), 237–248.

[Na] Nävert, U. A finite element method for convection diffusion problems, Thesis, Chalmers Univ. of Technology, 1982.

[RS] Rannacher, R., and Scott, R. Some optimal error estimates for piecewise linear finite element approximations, Math. Comp. 38 (1982), p 437–445.

[SF] Strang, G., and Fix, G. An analysis of the finite element method, Prentice-Hall, 1973.

[Sz] Szepessy, A. A streamline diffusion finite element method for conservation laws, Technical report, Math. Dept., Chalmers Univ. of Technology, 1988.

[T2] Temam, R. Problèmes Mathématiques en Plasticité, Gauthier-Villars, 1983.

[T1] Temam, R. Theory and numerical analysis of the Navier-Stokes equations, North Holland, 1977.

[Th] Thomée, V. Galerkin Finite Element Methods for Parabolic Problems, Lecture Notes in Mathematics 1054, Springer, 1984.

[W] Wendland, W. L. Boundary element methods and their asymtotic convergence, in Filippi, P. Theoretical acoustics and numerical techniques, CISM, Courses and Lectures 277, Springer 1983, 135–216.

[Z] Zienkiewicz. The Finite Element Method in Engineering Science, McGraw-Hill, 1971.

Index